10 Joint Workshop "Multidisciplinary Remote Sensing for Environmental Monitoring"

TC III WG III/2

ISPRS Archives Volume XLII-3/W7, 2019

Kyoto, Japan
12 – 14 March 2019

Editors:

M. Yoshimura
J. Susaki

ISBN: 978-1-5108-9790-8

Printed from e-media with permission by:

Curran Associates, Inc.
57 Morehouse Lane
Red Hook, NY 12571

CURRAN ASSOCIATES INC.
proceedings
.com

Some format issues inherent in the e-media version may also appear in this print version.

International Society for Photogrammetry and Remote Sensing (ISPRS)
c/o
Leibniz University Hannover
Institute of Photogrammetry and GeoInformation
Nienburger Str. 1
D-30167 Hannover, Germany

www.isprs.org

Additional copies of this publication are available from:

Curran Associates, Inc.
57 Morehouse Lane
Red Hook, NY 12571 USA
Phone: 845-758-0400
Fax: 845-758-2633
Email: curran@proceedings.com
Web: www.proceedings.com

TABLE OF CONTENTS

FEASIBILITY OF USING HYPERSPECTRAL REMOTE SENSING FOR ENVIRONMENTAL HEAVY METAL MONITORING 1
X. Chen, H. Lee, M. Lee

THIN ICE AREA EXTRACTION IN THE SEASONAL SEA ICE ZONES OF THE NORTHERN HEMISPHERE USING ASMR2 DATA 5
K. Cho, K. Miyao, K. Naoki

DEVELOPMENT OF PHENOLOGY BASED ALGORITHM FOR CROPLAND AND CROP TYPE MAPPING WITH MULTITEMPORAL LANDSAT IMAGE DATA - CASE STUDY IN THE NORTHWEST OF VIETNAM 10
N. D. Duong, N. M. Phuong, N. B. Thi

CHARACTERISTICS OF THE DEGREE OF GRADE IN GRADE-ADDED ROUGH SET FOR LAND COVER CLASSIFICATION 17
Y. Ishii, K. Iwao, T. Kinoshita

A FIT-FOR-PURPOSE ALGORITHM FOR ENVIRONMENTAL MONITORING BASED ON MAXIMUM LIKELIHOOD, SUPPORT VECTOR MACHINE AND RANDOM FOREST 23
A. Jamali

AGROFORESTRY TREE DENSITY ESTIMATION BASED ON HEMISPHERICAL PHOTOS & LANDSAT 8 OLI/TIRS IMAGE: A CASE STUDY AT CIDANAU WATERSHED, BANTEN-INDONESIA 31
R. N. Khairiah, L. B. Prasetyo, Y. Setiawan

ANALYSIS OF RELATIONSHIP BETWEEN VEGETATION DISTRIBUTION AND LAND PRICE USING MULTITEMPORAL DATA 38
K. Kumagai, H. Uematsu

GROUND PENETRATING RADAR DATA INTERPRETATION USING ELECTROMAGNETIC FIELD ANALYSIS FOR SEA ICE THICKNESS MEASUREMENT 43
M. Matsumoto, M. Yoshimura, K. Naoki, K. Cho, H. Wakabayashi

USING REMOTE SENSING DATA FOR CALCULATING THE COASTAL EROSION IN SOUTHERN THAILAND 47
W. Pantanahiran

MAPPING THE SPATIAL DISTRIBUTION OF COLOMBIA'S FOREST ABOVEGROUND BIOMASS USING SAR AND OPTICAL DATA 53
P. Rodríguez-Veiga, A. P. Barbosa-Herrera, J. S. Barreto-Silva, P. C. Bispo, E. Cabrera, C. Capachero, G. Galindo, Y. Gou, L. M. Moreno, V. Louis, P. Lozano, A. M. Pacheco-Pascagaza, I. P. Pachon-Cendales, J. F. Phillips-Bernal, J. Roberts, N. R. Salinas, L. Vergara, A. C. Zuluaga, H. Balzter

STUDY ON EXTRACTION METHODOLOGY OF RICE GROWTH CHARACTERISTICS IN FIELD USING TERRESTRIAL LASER MEASUREMENT DATA 57
K. Sakata, A. Rikimaru, Y. Yasunaga

SUMMARY OF NATURAL DISASTER RESPONSES BY THE ADVANCED LAND OBSERVING SATELLITE-2 (ALOS-2) 64
T. Tadono, M. Ohki, T. Abe

FLOODED AREA EXTRACTION OF RICE PADDY FIELD IN INDONESIA USING SENTINEL-1 SAR DATA 68
H. Wakabayashi, K. Motohashi, T. Kitagami, B. Tjahjono, S. Dewayani, D. Hidayat, C. Hongo

CHLOROPHYLL ESTIMATION OF LAKE WATER AND COASTAL WATER USING LANDSAT-8 AND SENTINEL-2A SATELLITE 72
S. Yadav, Y. Yamashiki, J. Susaki, Y. Yamashita, K. Ishikawa

ASSESSMENT OF CLOUD COVER CHARACTERISTICS OVER CALIBRATION TEST SITES USING MODIS CLOUD MASK PRODUCTS 78
H. Yamamoto, S. Tsuchida

Author Index

FEASIBILITY OF USING HYPERSPECTRAL REMOTE SENSING FOR ENVIRONMENTAL HEAVY METAL MONITORING

Xin-Yu Chen, Hui-Chun Lee, Mengshan Lee*

Department of Safety, Health and Environmental Engineering, National Kaohsiung University of Science and Technology, Kaohsiung, Taiwan

Commission III, WG III/10

KEY WORDS：Remote sensing; Hyperspectral data; Heavy metals; Non-destructive analysis; XRF; Pollution management strategy

ABSTRACT:

The use of optical properties as key parameters has been widely used in water quality monitoring, which accelerates the advances of remote sensing in the field of environmental monitoring. Current analytical methods for determining heavy metals in water include flame atomic absorption spectrometry (FAAS), atomic adsorption spectrophotometry (AAS) and inductively coupled plasma (ICP) spectroscopy, which typically require use of chemicals for sample processing and pretreatment as well as high capital input for analysis. Therefore, this study aims at investigating the potential of using non-destructive approaches for rapid water monitoring of heavy metal from green chemistry perspective. The proposed non-destructive sensing techniques include X-ray fluorescence spectrometer (XRF) and visible-near infrared spectroradiometer (VNIR). The former is an elemental analyser specifically for elements with relatively high atomic number, and the latter measures the reflectance or transmittance from samples. Heavy metals of lead (Pb), zinc (Zn) and copper (Cu) were selected as the target water constitutes in the study. The results from the analysis were then be used for determining a correlation model through chemometric approaches. Our results demonstrated that both of the target metals could be analysed via the proposed analytical methods. Reasonable agreements between the measurements from XRF and ICP were observed, whereas moderate correlations were perceived for simple linear regression model using spectral information from VNIR. Results from this study are expected to provide useful information on rapid identification of metal-polluting sources.

1. INTRODUCTION

Current conventional analytical methods for determining heavy metals in water include atomic absorption/emission spectroscopy (AAS), inductively coupled plasma mass spectrometry (ICP-MS) and cold vapour atomic fluorescence spectrometry (CV-AFS), which all require use of chemicals for sample processing and pre-treatment as well as high capital input for sampling and analysis. The demand for good quality, real-time, non-destructive and high resolution water quality information therefore has been growing in research areas of environmental monitoring, in order to comply with the concept of green chemistry.

The use of optical properties as key parameters has been widely used in environmental monitoring. Recent efforts were therefore made in the development of emerging analytical approach using non-destructive spectroscopic techniques, including laser induced breakdown spectrometry (LIBS), X-ray fluorescence (XRF) and visible-near infrared spectroradiometer (VNIR). Use of hyperspectral technique for determination of heavy metal in environmental samples were seldom reported, especially for water bodies. Rouillon et al. (2017) applied the XRF for measuring Mn, Zn and Pb at metal-contaminated sites. Dong et al. (2016) revealed the use of spectral data for estimating heavy metals in mining reclamation areas. Aldabaa et al. (2015) demonstrated the feasibility of using proximal (potable-XRF and VNIR) sensing for rapid soil salinity quantification; while other studies also shown comparable potential for using the abovementioned techniques for prediction of organic matter (Morona et al., 2017) and pH value (Sharma et al, 2014) in soil.

All of the proposed approaches are expected to be used as alternatives for conventional analytical techniques to provide reliable environmental monitoring results. But chemometric treatment, such as multi-linear regression, stepwise multi-linear regression, partial least squares regression, on the spectral data all played an important role in improving the accuracy on the prediction models (Gredilla et al., 2016; Dörnhöfer & Oppelt, 2016).

This study aims to explore the feasibility of using non-destructive spectrometer optical instruments, including high Spectrometer (VNIR) and potable X-ray fluorescence spectrometer (PXRF), for analysing the content of heavy metals in water bodies. Clarification on their applicability for field study is also addressed.

2. RESEARCH METHODS

2.1 Water sample preparation

The water samples evaluated in this study were prepared in our laboratory by diluting of stock solutions. Heavy metals of lead (Pb), zinc (Zn) and copper (Cu) were selected as the target water constitutes in the study. Range of concentration are 5-10 mg/L and 500-3000 mg/L for PXRF and VNIR, respectively.

2.2 Heavy metal content measurement

Two non-destructive spectrometer optical instruments, including high Spectrometer (VNIR, FieldSpec HandHeld 2) and potable X-ray fluorescence spectrometer (PXRF, Vanta M-Series), were selected in this study for estimating the content of heavy metals in water samples. A conventional chemical

* Corresponding author: mlee@nkust.edu.tw

https://doi.org/10.5194/isprs-archives-XLII-3-W7-1-2019

detection method of ICP-OES was used for comparison and elucidating the correlation between the approaches.

2.3 Experimental setup

Figure 1 reveals the experimental setup for the remote sensing technique using visible-near infrared spectroradiometer (VNIR). The sensing experiments were conducted in the light booth, which was intended to simulate the actual situation for field studies. The light booth enables the simulation of daylight and equipped with white interior as background reference. The water samples were stored in a quartz beaker during sensing measurements, in order to minimize the reflective interference from the beaker.

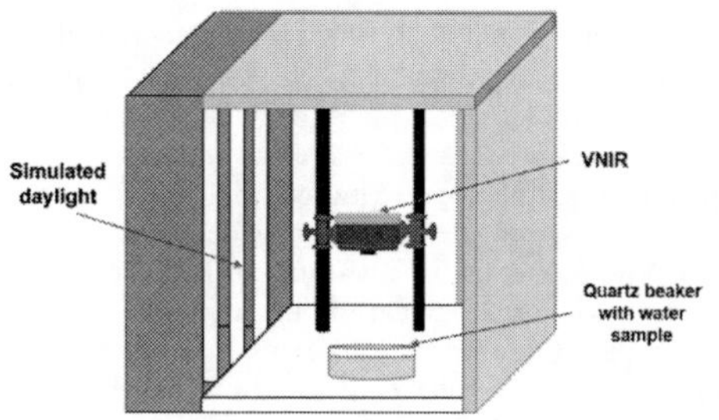

Figure 1. Setup of light booth for environmental simulation for sensing measurements

3. RESULTS AND DISCUSSION

3.1 Identification of spectral characteristics

Figure 2 shows the spectral data (from 800 nm to 1075 nm) for the studied heavy metal using the environmental simulation setup proposed in this study. The spectral curves without pretreatment were relatively smooth with no abnormal reflection peaks, which were directly used to identify the sensitive bands for the target metals. The sensitive bands were characterized as 984 nm, 968 nm and 1024 nm for Cu, Zn and Pb, respectively. Spectral characteristics on other sensitive bands for the studied metals are provided in Table 1. It was shown that most of the sensitive bands for the studied heavy metals fall in the range from 800 nm to 1075 nm, which were in line with the findings in Rostom et al.(2017).

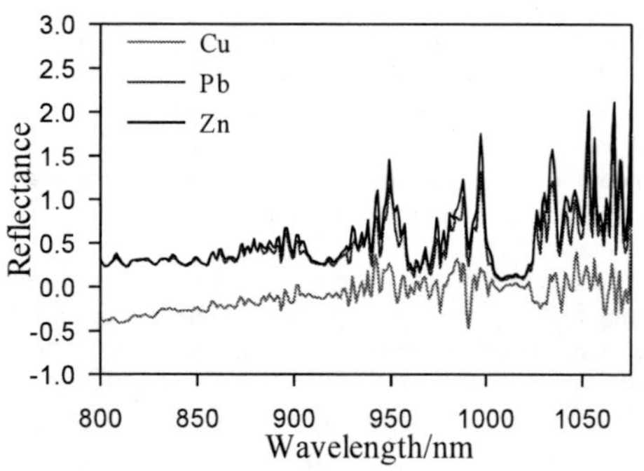

Figure 2. Spectral data for the studied metal elements (both of the samples are at 2000 mg/L)

Table 1. Spectral data for the studied metal elements

Metal	Spectral characteristics(nm)
Cu	946、984、988、997、1047、1065、1075
Pb	951、989、997、1024、1053、1066、1075
Zn	949、968、996、1034、1041、1069、1074

Figure 3 presents the spectra for the studied metal elements at different concentration from 500 mg/L to 3000 mg/L. It was observed that the reflectance on sensitive bands improved as increase of metal concentrations. Detail spectral data for the metals are provided in Table 2 to Table 4. All of the findings suggested the need of data pretreatment or select of appropriate chemometric approach to improve the accuracy and reliability on prediction model (Dörnhöfer & Oppelt, 2016)。

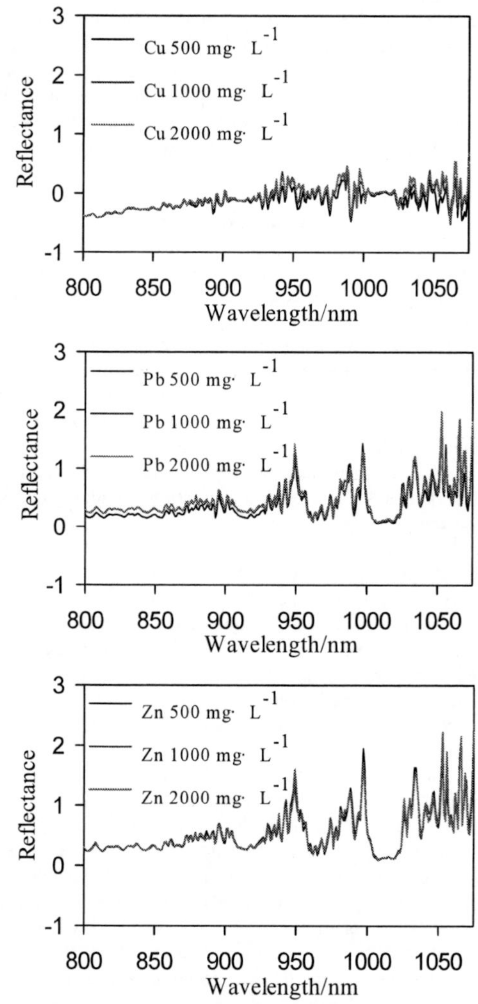

Figure 3. Spectra of the studied metal elements under different concentration (500 to 2000 mg/L)

Table 2. Reflectance characteristics at sensitive bands for Cu

Conc.(mg/L) / Spectral(nm)	500	1000	2000
946	0.23	0.19	0.29
984	0.20	0.28	0.27
988	0.33	0.28	0.47
997	0.13	0.21	0.42
1047	0.29	0.40	0.32
1065	0.25	0.27	0.54
1075	0.56	0.70	0.56

Table 3. Reflectance characteristics at sensitive bands for Pb

Conc.(mg/L) / Spectral(nm)	500	1000	2000
951	0.72	0.93	0.89
989	0.75	0.87	0.81
997	1.43	1.31	1.28
1024	0.22	0.27	0.19
1053	1.58	1.77	1.98
1066	1.58	1.60	1.85
1075	1.74	1.76	1.81

Table 4. Reflectance characteristics at sensitive bands for Zn

Conc.(mg/L) / Spectral(nm)	500	1000	2000
949	1.58	1.45	1.61
968	0.45	0.46	0.46
996	1.28	1.26	1.19
1034	1.64	1.57	1.56
1041	1.02	1.04	1.05
1069	1.37	1.45	1.55
1075	1.32	1.48	1.27

3.2 Spectral estimation results

Results from the spectral analysis were then be used for correlation analysis between the measured content and the spectral reflectance or elemental results.

Figure 3 shows the correlation of measured metal concentration in the validation samples (by ICP-OES) compared to the proposed non-destructive approach using PXRF. Significant correlation was obtained for the three studied metals, suggesting the high feasibility of using PXRF for detecting heavy metals in water samples. However, our results also validate the fact that the PXRF was not applicable for remote sensing application, as the intensity of fluorescence is inversely proportional to the distance between receptor and sample (Zhou et al., 2018).

Differently, the correlation between measured metal concentration and predicted results from non-destructive approach using VNIR were less significant but still at moderate levels, particularly for the Pb and Zn. The relatively higher correlation found for Cu could be attributed to the blue color characteristic of the solution, whereas the others remain clear. Future research efforts are therefore suggested to apply various multivariate calibrations or data mining techniques for improving the accuracy of prediction models.

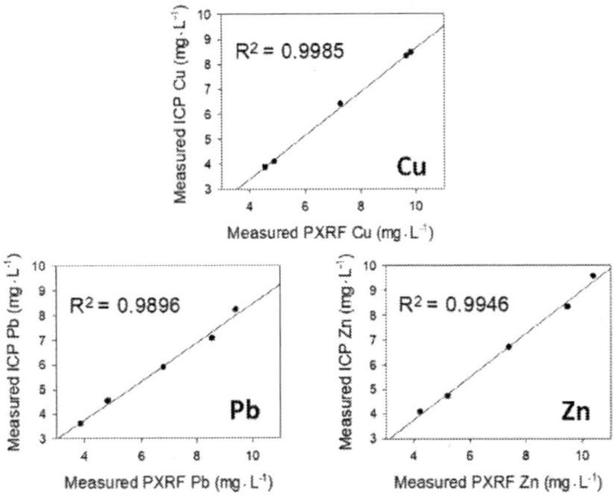

Figure 3. The measured metal concentration in the validation samples (by ICP-OES) compared to the proposed non-destructive approach using PXRF (at 8.9, 9.57, 89.94 keV for Cu, Zn, Pb, respectively)

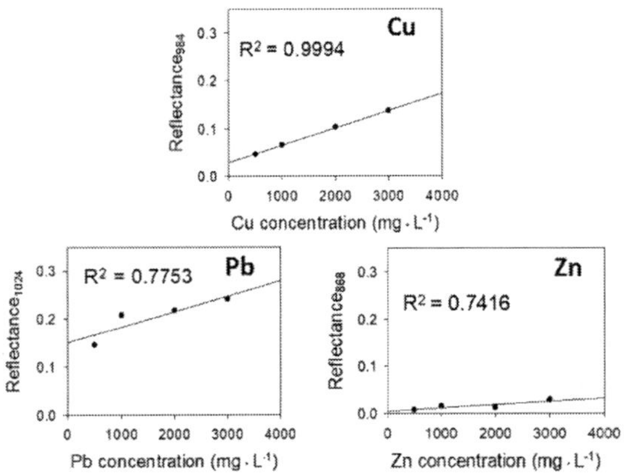

Figure 4. The measured metal concentration in the validation samples (by ICP-OES) compared to the proposed non-destructive approach using VNIR

4. CONCLUSIONS

The study demonstrated the use of non-destructive analytical methods, including PXRF and VNIR for quantification of metal contents (Cu, Pb, Zn) in water samples. Reasonable agreements between the measurements from validated samples and PXRF were observed, whereas moderate correlations were perceived for simple linear regression model using spectral information from VNIR. Results from this study are expected to provide useful information on rapid identification of metal-polluting sources.

ACKNOWLEDGEMENTS

The authors are grateful for the research funding from the Ministry of Science and Technology in Taiwan through project number 107-2218-E-992 -305.

REFERENCES

Aldabaa, A.A.A., Weindorf, D.C., Chakraborty, S., Sharma, A., Li, B. 2015. Combination of proximal and remote sensing methods for rapid soil salinity quantification. *Geaderma*, 239-240, 36-46.

Dong, J., Dai, W., Xu, J., Li, S. 2016. Spectral estimation model construction of heavy metals in mining reclamation areas. *International Journal of Environmental Research and Public Health*, 13, 640.

Gredilla, A., Fdez-Ortiz de Vallejuelo, S., Elejoste, N., de Diego, A., Madariaga, J.M. 2016. Non-destructive Spectroscopy combined with chemometrics as a tool for Green Chemical Analysis of environmental samples: A review. *TrAC Trends in Analytical Chemistry*, 76, 30-39.

Morona, F., dos Santos, F.R., Brinatti, A.M., Melquiades, F.L. 2017. Quick analysis of organic matter in soil by energy-dipersive X-ray fluorescence and multivariate analysis. *Applied Radiation and Isotopes*, 130, 13-20.

Rouillon, F., Taylor, M.P., Dong, C. 2017. Reducing risk and increasing confidence of decision making at a lower cost: In-situ pXRF assessment of metal-contaminated sites. *Environmental Pollution*, 229, 780-789.

Sharma, A., Weindorf, D.C., Man, T., Aldabaa, A.A.A., Chakraborty, S. 2014. Characterizing soils via portable X-ray fluorescence spectrometer: 3. Soil reaction (pH). *Geoderma*, 232-234, 141-147.

Zhou, S., Yuan, Z., Cheng, Q., Zhang, Z., Yang, J. 2018. Rapid in situ determination of heavy metal concentrations in polluted water via portable XRF: Using Cu and Pb as example. *Environmental Pollution*, 243, 1325-1333.

de la Mare, W., Ellis, N., Pascual, R., Tickell, S. 2012. An empirical model of water quality for use in rapid management strategy evaluation in Southeast Queensland, Australia. *Marine Pollution Bulletin*, 64(4), 704-711

Dörnhöfer, K., Oppelt, N. 2016. Remote sensing for lake research and monitoring – Recent advances. *Ecological Indicators*, 64, 105-122.

Rostom, N.G., Shalaby, A.A., Issa, Y.M., Afifi, A.A. 2017. Evaluation of Mariut Lake water quality using Hyperspectral Remote Sensing and laboratory works. *The Egyptian Journal of Remote Sensing and Space Science*, 20, S39-S48.

This contribution has been peer-reviewed.
https://doi.org/10.5194/isprs-archives-XLII-3-W7-1-2019

The International Archives of the Photogrammetry, Remote Sensing and Spatial Information Sciences, Volume XLII-3/W7, 2019
TC III WG III/2,10 Joint Workshop "Multidisciplinary Remote Sensing for Environmental Monitoring", 12–14 March 2019, Kyoto, Japan

THIN ICE AREA EXTRACTION IN THE SEASONAL SEA ICE ZONES OF THE NORTHERN HEMISPHERE USING ASMR2 DATA

K. Cho [1]*, K. Miyao [1], K. Naoki [1]

[1]Tokai University, 2-28-4, Tomigaya, Shibuya-ku, Tokyo, Japan, kohei.cho@tokai-u.jp

Commission III, WG III/2

KEY WORDS: sea ice, passive microwave radiometer, global warming, GCOM-W

ABSTRACT:

Sea ice has an important role of reflecting the solar radiation back into space. In addition, the heat flux of ice in thin ice areas is strongly affected by the ice thickness difference. Therefore, ice thickness is one of the most important parameters of sea ice. In our previous study, the authors have developed a thin ice area extraction algorithm using passive microwave radiometer AMSR2 for the Sea of Okhotsk. The basic idea of the algorithm is to use the brightness temperature scatter plots of AMSR2 19GHz polarization difference (V-H) vs 19GHz V polarization. The algorithm was also applicable to the Bering Sea, and could extract most of the thin ice areas. However, two problems have become clear. One was that some the thin ice areas were not well extracted, and the other was that some of the consolidated ice were mis-extracted as thin ice areas. In this study, the authors have improved the thin ice area extraction algorithm to solve these problems. By adjusting the parameters of the algorithm applied to the brightness temperature scatter plots of AMSR2 19GHz polarization difference (V-H) vs 19GHz V polarization, most of the thin ice areas were also well extracted in the Bering Sea. The authors also introduced an equation using the brightness temperatures difference of 89GHz vertical and horizontal polarization to reject the thin ice area misextracted over consolidated ice. By applying the above two methods to AMSR2 data, most of the thin ice areas in the Bering Sea were well extracted. The algorithm was also applied to the Gulf of St. Lawrence with good result. The thin ice area extracted data are planed to be approved by JAXA as a AMSR2 research product.

1. INTRODUCTION

Since 1978, passive microwave radiometers, including AMSR2 onboard GCOM-W satellite, have been continuously observing the earth for 40 years. The long-term sea ice extent derived from the passive microwave observation showed clear decline trend of the Arctic sea ice cover (Comiso, 2012, JAXA, 2012, NSIDC, 2018 etc.). The result is referred as an evidence of global warming in the Fifth Assessment Report of IPCC (2014). Usually, sea ice extent is calculated from sea ice concentration derived from brightness temperatures measured by passive microwave radiometers. Ice concentration is the most fundamental parameter of sea ice which can be calculated from brightness temperatures measured by passive microwave radiometers. There are number of sea ice concentration algorithms including NASA Team Algorithm (Cavarieli et al., 1984), Bootstrap Algorithm (Comiso, 1995) and ASI Algorithm (Svendsen et al. .1987).
Since the heat flux of ice is strongly affected by the ice thickness (Maykut, 1978), ice thickness is another important parameter of sea ice. However, the sea ice thickness information cannot be estimated from the sea ice concentration data. Studies on estimating ice thickness from the brightness temperature data acquired from passive microwave radiometers onboard satellites have been done in the past including those of Tateyama et al. (2002), Martin et al. (2005), and Tamura et al. (2007). However, the detailed validation of the accuracy of the estimated sea ice thickness is still on the way. Estimating ice thickness from passive microwave radiometer is not easy. The authors have been developing a method to extract thin ice area using brightness temperature scatter plots of AMSR2 19GHz polarization difference (V-H) vs 19GHz V polarization for the Sea of Okhotsk (Cho et. Al, 2012, 2014, 2015). In this study, "thin ice" is defined as the ice which thickness is around less than 30cm. Cho et al. (2011, 2012) has done detailed studies on comparing the in situ ice thickness measurement result with the data observed by optical sensors such as RSI on FORMOSAT-2 and MODIS on Aqua/Terra. The result suggested that if the ice thickness is less than 30cm, under the less snow cover condition, the ice thickness difference can be detected with optical sensors such as RSI and MODIS. In this study, the MODIS images are used as the reference of identifying thin ice areas, the possibility of extract thin ice area with AMSR2 data are validated by comparing with the MODIS images. The result of applying the thin ice area extraction algorithm, hereafter referred to as the Thin Ice Algorithm, to AMSR2 data in the seasonal sea ice zones of the northern hemisphere including the Sea of Okhotsk, Bering Sea, and Gulf of St. Lawrence are presented in this paper.

2. TEST SITES

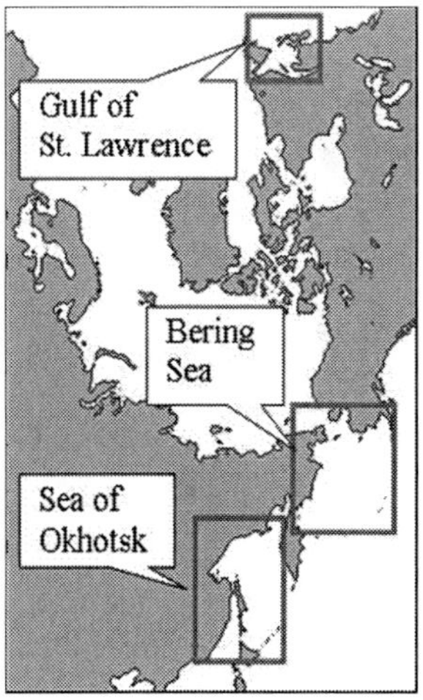

Figure 1 show the map of the test sites analyzed in this study which are the Sea of Okhotsk, the Bering sea and the Gulf of Saint Lawrence. All three are seasonal sea ice zones of the northern hemisphere. The Sea of Okhotsk is located at the north side of Hokkaido, Japan, surrounded by the Island of Sakhalin and eastern Siberian coast, Kamchatka Peninsula and Kuril Islands. The sea is one of the most southern seasonal sea ice zones in the northern hemisphere, and many thin

*Kohei Cho, Tokai University Research & Information Center, 2-28-4, Tomigaya, Shibuya-ku, Tokyo151-0063, Japan

ice areas can be found in the sea. The Bering Sea is located in the northernmost part of the Pacific Ocean, which is surrounded by the Siberia, the Kamchatka Peninsula, the Alaska Peninsula and the Aleutian Islands. The Bering Sea is connected to Arctic Ocean by the Bering Strait. The Gulf of St. Lawrence is a kind of inland sea located in eastern Canada. It is the outlet of North America's Great Lakes via the Saint Lawrence River into the Atlantic Ocean.

3. ANALYZED DATA

The brightness temperature data acquired from passive microwave radiometer AMSR2 onboard GCOM-W satellite were used in this study. GCOM-W was launched by JAXA in 2012 and AMSR2 has been observing the earth for over 6 years. Table 1 shows the specifications of AMSR2. The ice concentration data derived from AMSR2 data using Bootstrap Algorithm (Comiso, 2009) were also used in this study. In order to identify thin ice areas, data collected by optical sensor MODIS onboard Aqua satellite were used as reference. Table 3 show the specifications of MODIS. As for MODIS, only the Band 1 and 2 which have the highest spatial resolution of 250m were used in this study. Under the cloud free condition, detailed distribution of sea ice can be observed from MODIS images. Since Aqua and GCOM-W are in the same orbital "track" under the frame work of the NASA's A-Train (NASA, 2012), the constellation of satellites, MODIS onboard Aqua observed the same area four minutes after the observation of AMSR2 onboard GCOM-W. Therefore, MODIS data is one of the most effective validation data for AMSR2 data.

Table 1. Specifications of AMSR2

Frequency (polarization)	IFOV	Swath	Incident angle
7GHzV,H)	35×62 km	1450 km	55 deg
11GHz (V,H)	24×42 km		
19GHz(V,H)	14×22 km		
24GHz(V,H)	15×26 km		
36GHz(V,H)	7×12 km		
89GHz(V,H)	3×5 km		

Table 2. Specifications of MODIS

Band	Wavelength	IFOV	Swath
1	0.620-0.670 μm	250 m	2330 km
2	0.841-0.876 μm		

4. RESEARCH METHOD

4.1 Sample Area selection

Figure 2 show the comparison of simultaneously collected AMSR2 ice concentration image and MODIS image of the Sea of Okhotsk taken on February 27, 2013. The clear distribution of sea ice can be identified from the AMSR2 ice concentration image as shown on Figure 2(a).

However, it is difficult to identify ice thickness differences or thin ice areas from the image. On the other hand, more detailed sea ice distributions can be observed in the color composite image of MODIS(Band 1 to blue and red, Band 2 to green) as shown on Figure 2(b). In our thin ice algorithm, we first select sample areas of thin sea ice, big ice floe, open water and mixed sea ice are selected in this study as sown on Figure 2 and 3.

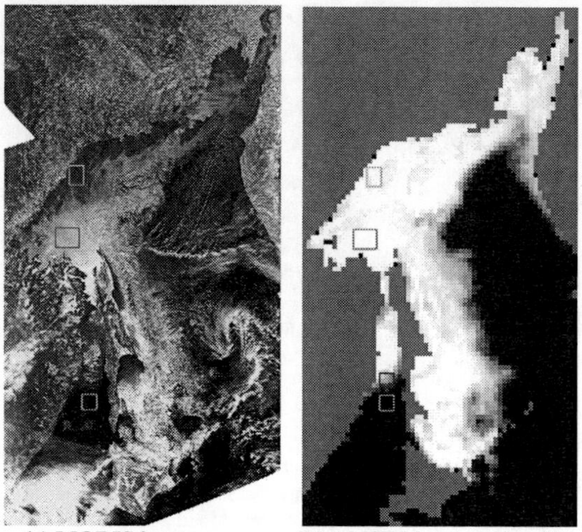

(a) MODIS image (b) AMSR2 ice concentration

Figure 2. Comparison of AMSR2 and MODIS images. (Sea of Okhotsk, February 27,2013)

(a)Big ice floe (b)Thin ice (c)Mixed ice (d)Open water

Figure 3. Sample area of different ice types extracted from MODIS image (Sea of Okhotsk, Feb. 27, 2013)

4.2 Thin Ice Algorithm

Figure 4 shows the scatter plot of AMSR2 19GHz V versus 19GHz (V-H) of the Sea of Okhotsk observed on February27, 2013. In this scatter plot, ▲ represents thin ice and ■ represents big ice floe. The basic idea of our Thin Ice Algorithm is to extract the thin ice area by applying the following two equations to the brightness temperatures (Tb) of 19GHz Vertical(V) and Horizontal(H) polarization of AMSR2.

(Tb19GHzV)>T1 (1)

(Tb19GHzV−Tb19GHzH) > -Tb19GHzV + T2 (2)

where Tb19GHzH: Tb of AMSR2 19GHz H polarization
Tb19GHzV: Tb of AMSR2 19GHz V-polarization
T1, T2: parameters adjusted to particular sea ice zone.

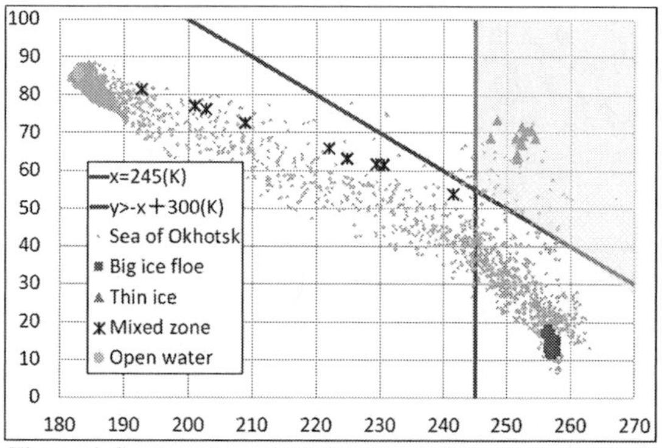

Figure 4. Scatter plots of (19GHzV – 19GHzH) Vs 19GHzV Polarization (Sea of Okhotsk, Feb. 27, 2013)

This contribution has been peer-reviewed.
https://doi.org/10.5194/isprs-archives-XLII-3-W7-5-2019 | © Authors 2019. CC BY 4.0 License.

It is impossible to identify ice thickness difference in the low ice concentration areas. The equation (1) is used to extract sea ice area with 80% or higher sea ice concentration. In other words, our target is thin ice area which concentration is higher than 80%/ The microwave brightness temperature of water is much lower in H polarization than in that of V polarization. Since thin ice areas are rather wet, the microwave brightness temperature of thin ice areas become much lower in H polarization than in that of V polarization. On the other hand, the microwave brightness temperature of consolidated ice does not show big difference between V and H polarization. Considering these characteristics, the authors have introduced equation (2) for extracting thin ice area.

5. EXTRACTED RESULT

5.1 Sea of Okhotsk

Firstly, the authors have applied the Thin Ice Algorithm to AMSR2 data of the Sea of Okhotsk. Figure 5(a) shows the AMSR2 sea ice concentration image of February 27, 2013. The cyan areas in the image show the "thin ice areas" extracted using AMSR2 data using equations (1) and (2). T1=245K and T2=300K are specified for the Sea of Okhotsk. The extracted areas were overlaid on the simultaneously collected MODIS image for evaluation as shown on Figure 5 (b). It shows that not all but most of the thin ice areas which are appearing in dark purple in the MODIS image are extracted with the proposed method.

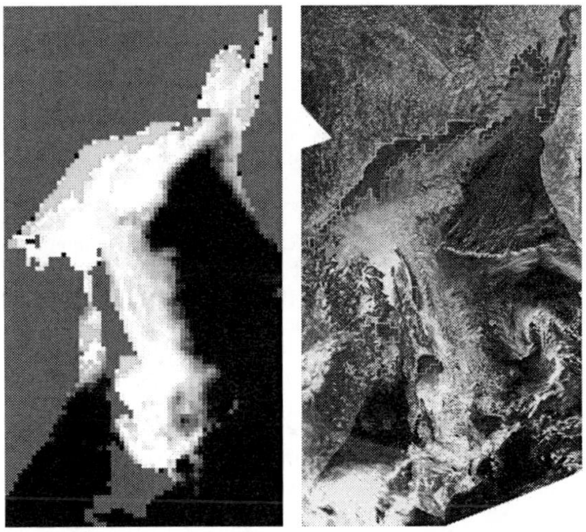

(a)AMSR2 image (b) MODIS image

Figure 5. Thin ice area extraction result (Cyan: extracted area) (Sea of Okhotsk, Feb. 27, 2013)

5.2 Bering Sea

The authors have applied the Thin Ice Algorithm to several scenes of AMSR2 data for the Bering Sea observed on March 19, 2016. Figure 6 shows the scatter plot of AMSR2 19GHz V versus 19GHz (V-H) of the Bering Sea observed on March 19, 2016. The blue meshed area represents the area to be extracted as thin ice area with equation (1) and (2). It is clear that the un-extracted sea ice areas(■) are distributed outside of the blue meshed area. Therefore, the authors have changes parameter T1 of equation (1) from 245K to235K. No change to the parameter T2=300K. In order to reject the effect of mis-extracting some of the big ice floe as thin ice areas, the authors have added the following equation to the algorithm(see .Miyao et al, 2018)

$$(Tb89GHzV - Tb89GHzH) > 20K \quad (3)$$

where Tb89GHzV: Tb of AMSR2 89GHz V polarization
Tb89GHzH: Tb of AMSR2 89GHz H polarization

Figure 7 show the extracted thin ice areas overlaid on the AMSR2 sea ice concentration image and MODIS image of the Bering Sea observed on March 19, 2016.

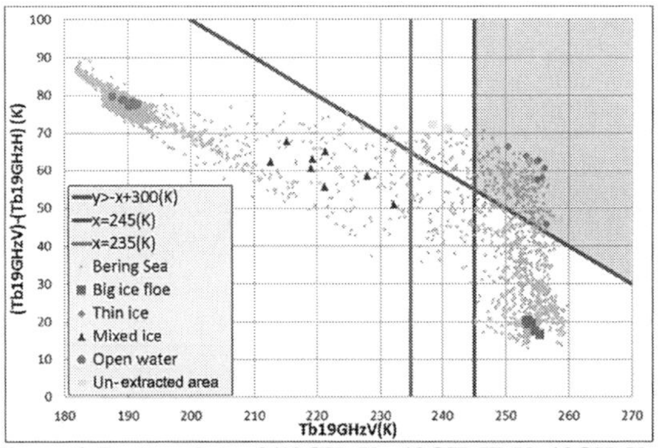

Figure 6. Scatter plots of (19GHzV – 19GHzH) Vs 19GHzV Polarization (Bering Sea, Mar. 19, 2016)

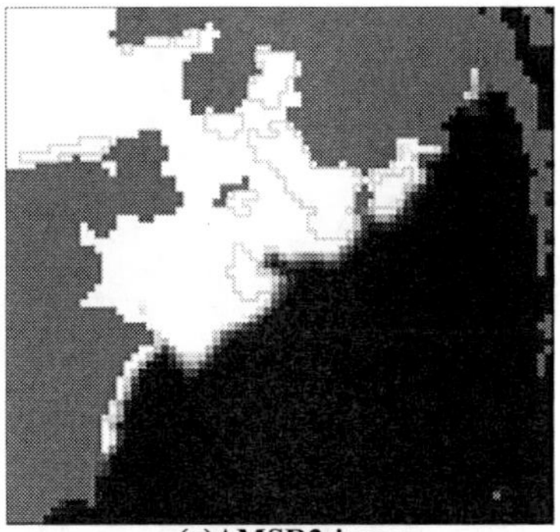

(a)AMSR2 image

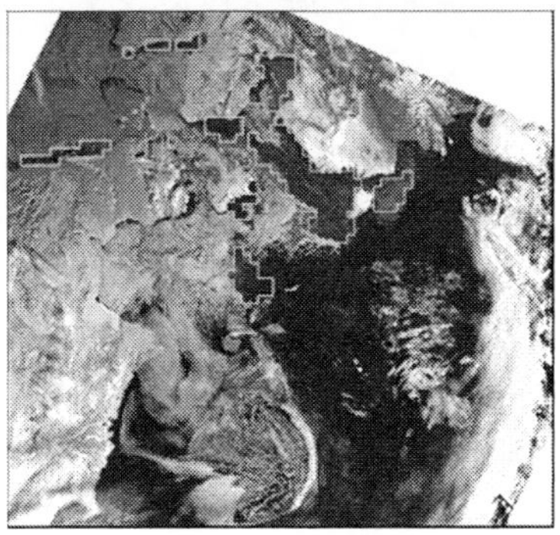

(b) MODIS image

Figure 7. Thin ice area extraction result (Cyan: extracted area) (Bering Sea, Feb. 10, 2014)

5.3 Gulf of St. Lawrence

Figure 8 show the scatter plot of AMSR2 19GHz V versus 19GHz (V-H) of the Gulf of St. Lawrence observed on March 1, 2015. The distribution of the data also suggests the possibility of applying the same algorithm to the Gulf of St. Lawrence. Just same with Bering Sea, the parameter T1 of equation (1) was changed from 245K to235K. No change to the parameter T2=300K. Figure 9 show the extracted thin ice areas overlaid on the AMSR2 sea ice concentration image and MODIS image of the Bering Sea observed on March 1, 2015.

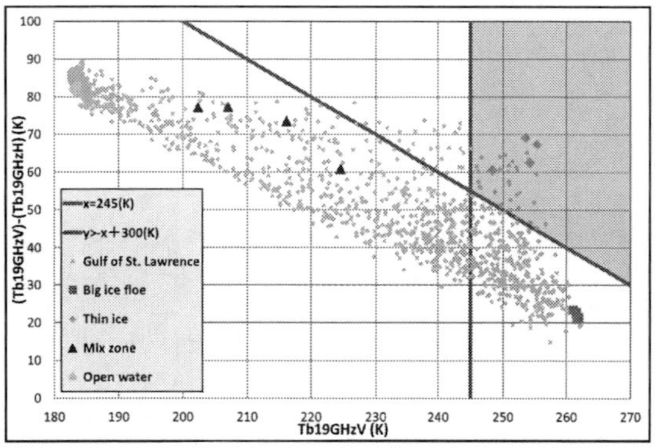

Figure 8. Scatter plots of (19GHzV – 19GHzH) Vs 19GHzV Polarization (Gulf of St. Lawrence, Mar. 1, 2015)

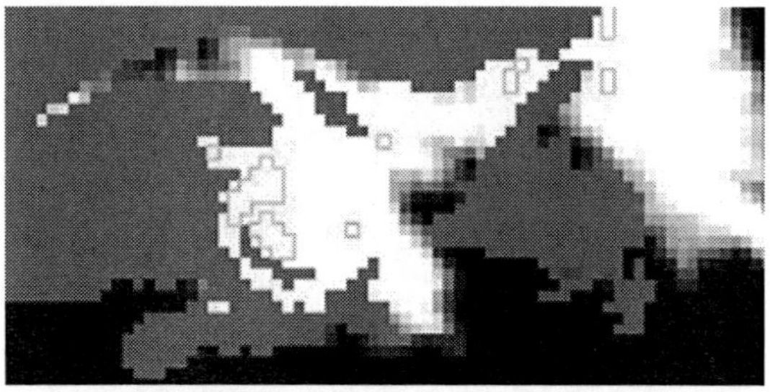

(a)AMSR2 image

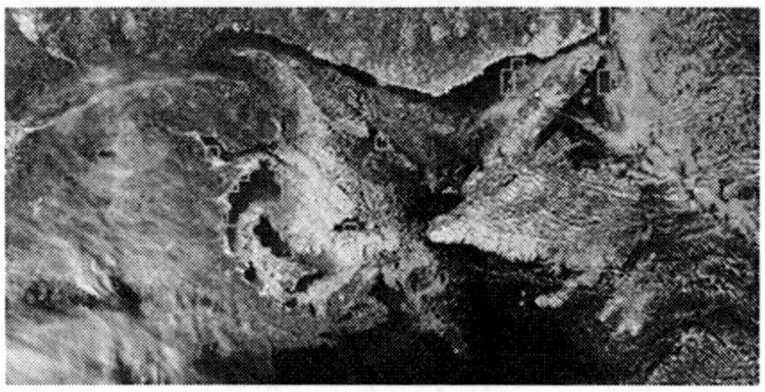

(b) MODIS image

Figure 9. Thin ice area extraction result (Cyan: extracted area) (Gulf of St. Lawrence, Mar. 1, 2015)

6. CONCLUSION

In this study, authors have applied the AMSR2 Thin Ice Algorithm which was originally developed for the Sea of Okhotsk also to the Bering Sea and the Gulf of St. Laurence. The extracted thin sea ice areas were validated by comparing with simultaneously collected MODIS images. The authors have analyzed around 10 scenes for the Sea of Okhotsk, the Bering Sea and the Gulf of St. Laurence. The most of the thin ice areas identified in MODIS images were well extracted from AMSR2 data by applying the algorithm. The result suggests that some tuning of the parameters may improve the accuracy of the algorithm when applying the algorithm to the other sea ice zones of the Northern Hemisphere. JAXA has decided to produce the thin ice product from AMSR2 data using the Thin Ice Algorithm as the research product of AMSR2.

ACKNOWLEDGEMENTS

This study was supported by JAXA under the framework of GCOM-W Project. The authors would like to thank JAXA for their kind support.

REFERENCES

Comiso, J., 2012, Large Decadal Decline of the Arctic Multiyear Ice Cover, Journal of Climate, Vol. 25, pp.1176-1193.

JAXA, 2012, Arctic Sea Ice Observation Data Analysis Results, Press Release, http://global.jaxa.jp/press/2012/08/20120825_arctic_sea_e.html.

NSIDC, 2018, Arctic sea ice maximum at second lowest in the satellite record, http://nsidc.org/arcticseaicenews/

IPCC. 2014. "Summary for Policymakers." In Climate Change 2013: The Physical Basis. Contribution of Working Group I to the Fifth Assessment Report of the Intergovernmental Panel on Climate Change, edited by T.F. Stocker, D. Qin, G.K. Plattner, et al., Cambridge: Cambridge University Press.

Cavalieri, D. J. and P. Gloersen, 1984, Determination of sea ice parameters with the NIMBUS 7 SMMR, J. Geophys. Res., Vol.89, pp.5355-5369.

Comiso, J. C., 1995 ,SSM/I Sea Ice Concentrations Using the Bootstrap Algorithm", NASA Reference Publication 1380, Maryland, NASA Center for AeroSpace Information.

Svendsen, E., C. Matzler & T. C. Grenfell, 1987, A model for retrieving total sea ice concentration from a spaceborne dual-polarized passive microwave instrument operating near 90 GHz, International Journal of Remote Sensing, Vol.8, No.10, pp.1479-1487

Maykut, G. A., 1978, Energy exchange over young sea ice in the central arctic, JGR, Vol.83, pp.3646-3658

Tateyama K., H. Enomoto,, T. Toyota. and S. Uto, 2002, Sea ice thickness estimated from passive microwave radiometers, Polar Meteorol. Glaciol., National Institute of Polar Research, Vol. 16, pp.15-31.

Martin, S. and R. Drucker, 2005, Improvements in the estimates of ice thickness and production in the Chukchi Sea polynyas derived from AMSR-E, GRL, Vol. 32, L05505

Tamura T., K. I. Ohshima, T. Markus, D. J. Cavalieri, S. Nihashi, N. Hirasawa, 2007, Estimation of Thin Ice Thickness and Detection of Fast Ice from SSM/I Data in the Antarctic Ocean, Journal of Atmospheric and Oceanic Technology, Vol. 24, pp.1757-1772.

Cho, K., Y. Mochizuki, Y. Yoshida, 2012, Thin ice area extraction using AMSR-E data in the Sea of Okhotsk, Proceedings of the 33rd Asian Conference on Remote Sensing, TS-E6-3, pp.1-6.

Tokutsu Y., K. Cho, 2014, Thin Ice Area Extraction Algorithm Using AMSR2 Data for the Sea of Okhotsk, Proceedings of the 35th Asian Conference on Remote Sensing, OS-101, pp.1-6.

Sato, Y., K. Naoki, K. Cho, A Study on Extracting the Trend of Then Ice Distribution in the Sea of Okhotsk Using ASMR-E and AMSR2 Data, Proceedings of the 36th Asian Conference on Remote Sensing, TUP1.49, pp.1-6. October 2015.

Cho K., Y. Mochiduki, Y. Yoshida, M. Nakayama, K. Naoki, C. F. Chen, 2011, Thin ice thickness monitoring with FORMOSAT-2 RSI data, Proceedings of the 32nd Asian Conference on Remote Sensing, TS1-2, pp.1-8

Cho K., Y. Mochizuki, Y. Yoshida, H. Shimoda and C. F. CHEN, 2012, A study on extracting thin sea ice area from space, International Archives of the Photogrammetry, Remote Sensing and Spatial Information Sciences, Vol. XXXIX-B8, pp.561-566.

Comiso, J. C., 2009, Enhanced Sea Ice Concentrations and Ice Extent from AMSR-E Data, Journal of the Remote Sensing Society of Japan, Vol.29, No.1, pp.199-215.

Miyao, K., K. Naoki, K. Cho, Upgrading thin ice area extraction algorithm using AMSR2 data, Proceedings of the 39th Asian Conference on Remote Sensing, ABA469, pp.1-10. October 2018.

NASA, 2012, https://atrain.nasa.gov/

This contribution has been peer-reviewed.
https://doi.org/10.5194/isprs-archives-XLII-3-W7-5-2019

DEVELOPMENT OF PHENOLOGY BASED ALGORITHM FOR CROPLAND AND CROP TYPE MAPPING WITH MULTITEMPORAL LANDSAT IMAGE DATA - CASE STUDY IN THE NORTHWEST OF VIETNAM

N. D. Duong[1*], N. M. Phuong[2], N. B. Thi[3]

[1]Institute of Geography, Vietnam Academy of Science and Technology, 18 Hoang Quoc Viet, Cau Giay, Hanoi, Vietnam – duongnd@geoifo.com.vn
[2] World Agroforestry Centre, Vietnam office, 249A Thuy Khue, Tay Ho, Hanoi, Vietnam - n.maiphuong@cgiar.org
[3]National Centre for Technological Progress, 25 Le Thanh Tong, Hoan Kiem, Hanoi, Vietnam - ngbthi@gmail.com

Commission III, WG III/10

KEY WORDS: Phenology, Cropland, Crop type, Simplified spectral pattern, Landsat, Multitemporal image data

ABSTRACT:

Cropland mapping is very important for food security, policy development, land use planning, and environmental protection. Scientists have developed methods and techniques for cropland mapping with remote sensing image data. Both single date and multitemporal image data are used for generation of cropland and crop type maps. Multitemporal image data has advantages over single date image data from reliability and accuracy point of view because multitemporal image data allows to eliminate seasonality of vegetation. In this paper, the authors present new algorithm for cropland and crop type mapping with multitemporal Landsat image data. The algorithm requires for analysis of all Landsat scenes observed during one year and if needed, scenes in some years back to compensate clouds and cloud shadows. Phenology of land cover is constructed based on six bimonthly cloud free land covers that were automatically classified using the selected scenes. By grouping land covers within two months to six land covers of periods January-February, March-April, May-June, July-August, September-October, and November-December we create six bimonthly cloud free land covers that formulate a database for mapping cropland and crop types. By analysis of 50 Landsat scenes of path/row number 128/45 (northwest of Vietnam) observed mainly from 2017, 2016 and 2015 we success to map upland cropland and 14 crop types with area ranging from 145,143 ha to 3,373 ha per crop type. The study pointed out that phenology characterized by six bimonthly land covers is acceptable to identify cropland distribution and some specific crop types. For better results, apparently we need higher temporal resolution of image data. Due to uncertainty of the atmosphere, it is almost impossible to rely only on optical remote sensing data to achieve high temporality of data so application of multitemporal SAR data could be a way to overcome this obstacle.

1. INTRODUCTION

Cropland mapping is very important for food security, policy development, land use planning and environmental protection (Csillik and Belgiu, 2017; Shao et al., 2009; Xiong et al., 2017). In conventional way, annual cropland maps are established by using official statistics together with land use map and field survey (Foerster et al., 2012; Victoria et al., 2012). Remote sensing image data, due to its large area coverage, periodically repeated and multispectral observation, has been applied in a multitude of different mapping areas, ranging from agriculture, forestry, land cover, land use, including cropland and crop type mapping. Both coarse and medium spatial resolution image data are used to generate cropland maps in national wide or continental scale (Waldner et al., 2017; Xiong et al., 2017; Zhao et al., 2017; Zhong et al., 2016).

Methodology for cropland mapping using remote sensing data has been developed for years (Simonetti et al., 2015; Xiong et al., 2017; Zhong et al., 2016). In general, adopted methods and techniques include phenology based algorithms (Chen et al., 2018; Dong et al., 2016; Foerster et al., 2012; Simonetti et al., 2015; Zhao et al., 2017; Zhong et al., 2016), classification decision tree (Deng and Wu, 2013), support vector machine (Brenning et al., 2006), random forest (Tatsumi et al., 2015), and others. Data used in these methods is either single date or multitemporal. Selection of single date images for cropland mapping depends on status of the vegetation and weather conditions. Usually, single date image is required to be taken at peak growing season and not affected by clouds or cloud shadows. Cropland maps generated by using single date image, however, involve uncertainties and inaccuracy caused by vegetation seasonality (Simonetti et al., 2015, 2014). Multitemporal image data has many advantages over single date image data in term of elimination of seasonality of vegetation. Phenology of cropland and crop types is characterized, mostly, by using time series of normalized vegetation index (NDVI) or the enhanced vegetation index (EVI). The use of vegetation index, nevertheless, possesses errors in the time series, such as sensor problems or cloud contamination not screened out by the fitting procedure, could lead to the incorrect identification of croplands (Victoria et al., 2012). Little study applied phenology characterized by temporal profile of land cover for accurate land cover or crop mapping.

In this paper, we present a development of a new algorithm for cropland and crop type mapping with Landsat multitemporal image data. Unlike conventional approach where time series of vegetation index is used to characterize phenology of vegetation, here we use temporal profile of land cover (variations of land cover type throughout a year) to describe growing cycle of field crops and then use this profile to classify cropland and crop types. We assume that there are very few cases when cloud free Landsat images are available for both dry and rainy seasons so we need

* Corresponding author

https://doi.org/10.5194/isprs-archives-XLII-3-W7-11-2019

to use more Landsat scenes for every two months to composite cloud free Landsat image and then to produce bimonthly cloud free land cover. A set of six bimonthly land covers is a kernel for development of land cover profile which is basis for crop land and crop type mapping. The study was tested in northwest of Vietnam where upland shifting cultivation is dominant form of agricultural production. We success to generate upland cropland map and map of 14 crop types with area ranging from 145,143 ha to 3,373 ha per crop type.

2. STUDY AREA

The study area is located in the northwest of Vietnam covered by Landsat scene with path/row number 128/045 (Fig. 1). The selected Landsat scene covers parts of Yen Bai, Son La and Dien Bien provinces of Vietnam where live about 30 different ethnic groups with majority of Thai, Kinh, H'mong, Muong and Dao peoples. The area is mountainous, remote and livelihoods of ethnic groups were dependent upon shifting cultivation practices (Wezel et al., 2006). The shifting cultivation practices are characterized by the clearing of primary or secondary forest to cultivate upland crops. The cropland landscape here is highly fragmented and heterogeneous with relatively small patches. This agricultural practices leads to deforestation and increase of soil erosion. Topography of the study area is rugged with Hoang Lien Son range where locates mount Fansipan - the highest mountain of Indochina peninsula with the altitude of 3,143m above sea level. Climate here is divided to two distinct season with rainy season starting from May to November. Vegetation is dominated by evergreen forest. Upland cultivation includes major crops as maize, sugarcane, upland rice, and cassava.

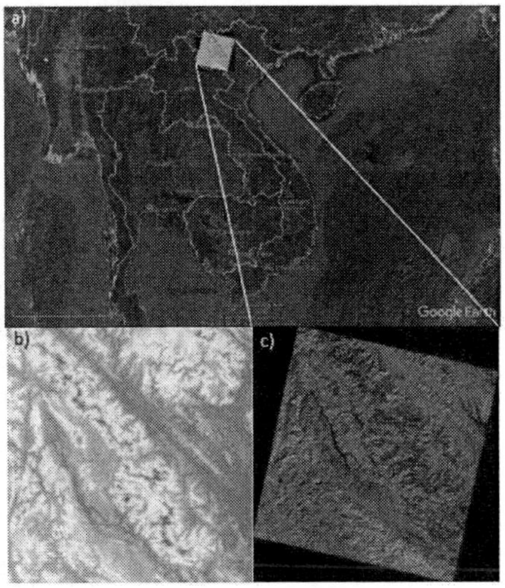

Figure 1. Study area; a) Location of study site in Vietnam; b) Digital elevation model; c) False colour composite of Landsat 8 RGB:654 of Dec. 20, 2017.

3. MATERIALS AND METHODS

3.1 Data preparation

Landsat image data was downloaded from United State Geological Service (USGS) website https://earthexplorer.usgs.gov/. Totally, 50 Landsat scenes with cloud coverage smaller than 80 % was selected. Despite the fact that baseline year for cropland mapping is 2017, we need to use Landsat scenes for the same months of 2016 and 2015 to compensate cloud cover in 2017 scenes. For very high mountain area with heavy cloudiness, we used some scenes from older years as 2006, 2009 and so to achieve cloud free dataset. This operation, in fact, does not severely influence the quality of the land cover since there are almost no changes over there due to inaccessibility. Table 1 shows list of 50 Landsat scenes used for this study.

Months	ID of Used Landsat Scenes
January-February	LC08_L1TP_128045_20150113_T1
	LC08_L1TP_128045_20150129_T1
	LC08_L1TP_128045_20170219_T1
	LE07_L1TP_128045_20060128_T1
	LE07_L1TP_128045_20070131_T1
	LE07_L1TP_128045_20140102_T1
	LE07_L1TP_128045_20140118_T1
	LE07_L1TP_128045_20160209_T1
	LT05_L1TP_128045_20090112_T1
	LT05_L1TP_128045_20090213_T1
March-April	LC08_L1TP_128045_20140331_T1
	LC08_L1TP_128045_20150318_T1
	LC08_L1TP_128045_20150403_T1
	LC08_L1TP_128045_20150419_T1
	LC08_L1TP_128045_20170323_T1
	LC08_L1TP_128045_20170408_T1
	LE07_L1TP_128045_20120402_T1
	LE07_L1TP_128045_20130320_T1
	LT05_L1TP_128045_20060410_T1
May-June	LC08_L1TP_128045_20130616_T1
	LC08_L1TP_128045_20140603_T2
	LC08_L1TP_128045_20150505_T1
	LC08_L1TP_128045_20160507_T1
	LC08_L1TP_128045_20160624_T1
	LC08_L1TP_128045_20170526_T1
	LE07_L1GT_128045_20170502_T2
	LE07_L1TP_128045_20070507_T2
	LE07_L1TP_128045_20110518_T2
	LE07_L1TP_128045_20120504_T2
	LE07_L1TP_128045_20150529_T1
July-August	LC08_L1TP_128045_20150708_T1
	LC08_L1TP_128045_20150809_T1
	LC08_L1TP_128045_20160710_T1
	LC08_L1TP_128045_20160726_T1
	LC08_L1TP_128045_20170729_T1
	LE07_L1TP_128045_20170822_T1
	LT05_L1TP_128045_19900804_T1
September-Otober	LC08_L1TP_128045_20140907_T1
	LC08_L1TP_128045_20141009_T1
	LC08_L1TP_128045_20151028_T1
	LC08_L1TP_128045_20160928_T1
	LC08_L1TP_128045_20171001_T1
	LE07_L1TP_128045_20101022_T1
	LT05_L1TP_128045_20070920_T1
November-December	LC08_L1TP_128045_20141126_T1
	LC08_L1TP_128045_20171220_T1
	LE07_L1TP_128045_20131217_T1
	LE07_L1TP_128045_20141220_T1
	LE07_L1TP_128045_20151223_T1
	LE07_L1TP_128045_20161209_T1

Table 1. List of fifty Landsat scenes used to generate bimonthly cloud free land covers of 2017

All downloaded Landsat scenes are of level-1 processing and before classification of land cover, digital number DN was converted to top of atmosphere (TOA) reflectance. Figure 2 show example of complexity of cloud cover of the study area during January and February of year 2017. This complexity explains reason for usage of more Landsat scenes to compensate cloud pixels to achieve cloud free land cover.

3.2 Algorithm development

Cropland and crop type mapping is very challenge due to the complexity of agricultural cultivation system. High temporality of remotely sensed data is one of key requirement for success of cropland and crop type mapping. Landsat image data has revisit observation of every 16 days. In cloud free atmosphere, this revisit frequency is good enough to observe different growing stage of crop. However, due to cloud contamination, this revisit time is not sufficient to obtain cloud free image data so we need more observation to compensate cloudy pixels.

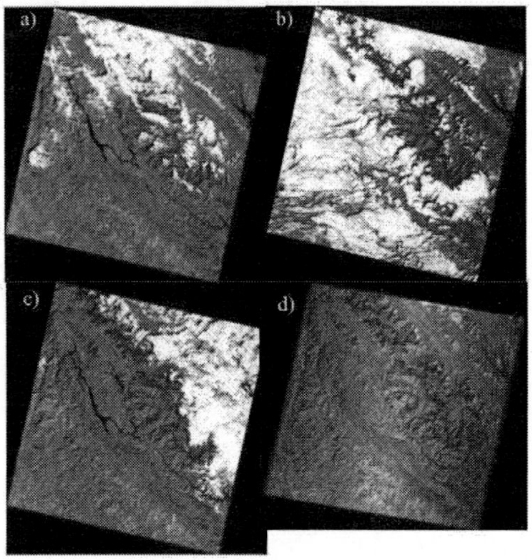

Figure 2. Landsat scenes show complexity of cloud cover in the study area for months of January and February of year 2017. Colour composite is RGB:654 for Landsat 8 and RGB:543 for Landsat 7. Observation date is Jan. 13, 2015 for a); Jan. 29, 2015 for b); Feb. 19, 2017 for c) and Jan. 28, 2006 for d).

Our proposed algorithm is shown in Fig. 3. The algorithm is divided to two main stages. In the first stage, a dataset of six bimonthly cloud free land covers is created. In the second stage, a phenological pattern database is established with the six bimonthly cloud free land covers. The database is organized statistically to make ease of identification of unique phenological patterns. Such developed database is used further for simulation of cropland and crop type distribution using phenological patterns. In the right side of Fig. 3 we present complementally flowchart of automated classification of land cover based on application of simplified spectral pattern (SSP) concept (Duong, 2018, 2016, Duong et al., 2017, 2014). For automated classification of land cover, we developed a database of simplified spectral patterns representing eight land cover categories: water, wetland, close forest, open forest, brown vegetation, sparse vegetation, dry bare land and moist bare land. The classification is based on comparison of SSP computed for every pixel of image to SSPs in the database and label the pixel by class of identical SSP.

3.2.1 Bimonthly cloud free land cover classification

Every selected Landsat scene will be automatically classified to obtain land cover with eight major classes: water, wetland, close forest, open forest, brown vegetation, sparse vegetation, dry bare land and moist bare land. Cloudy pixels are masked by using Band quality assessment band (BQA) which is provided in every Landsat Collection 1 level 1 data. We use cloud confidence of 75 percent to detect cloudy pixels.

We divide one-year time to six periods of two month clusters: January-February, March-April, May-June, July-August, September-October and November-December. Then we combine land covers within each period together to obtain cloud free land cover. During this combination, cloud free pixel originating from newest observation date will have higher priority than pixel from older data to be included in the final land cover (replacing cloudy pixels). Figure 4 displays six bimonthly cloud free land covers of the study area. When pixels from two land cover are cloud free at the same time, combination algorithm prefers pixel with less vegetation (bare land) to be included in final result.

3.2.2 Establishment of land cover database

In order to map efficiently cropland and crop types, we firstly create a table of all phenological patterns and then apply a phenological model to map cropland and crop types. Phenological model allows selection of phenological conditions following local ecological and cropping practices to classify pixels to particular crop type. For example, to map upland cropland, we select phenological patterns, which have bare land and vegetation in at least one period while no land cover class in any period is water or wetland. Phenological patterns in the database are sorted in descending order by number of pixels. The pattern with highest number of pixels is put on the first position in the sorting list. The number of phenological patterns is often very large so we focus on patterns on top of the sorting list only. These patterns are more meaningful than those with small number of pixels.

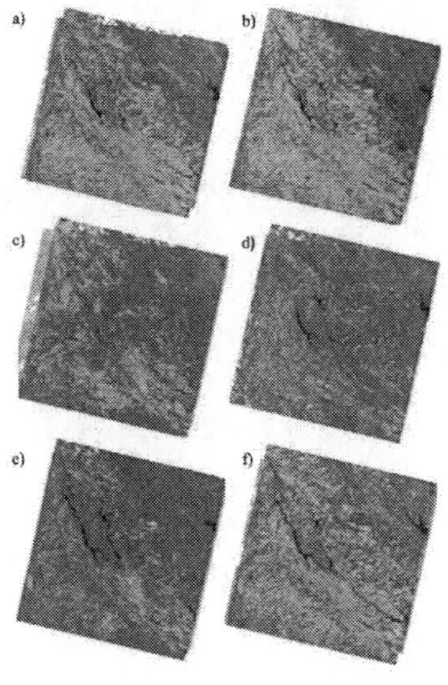

Figure 4. The six bimonthly cloud free land covers of year 2017 for periods: a) January-February, b) March-April, c) May-June, d) July-August, e) September-October, and f) November-December. Dark Green: forest; Yellow green: open forest; Blue: water; Orange: bare land.

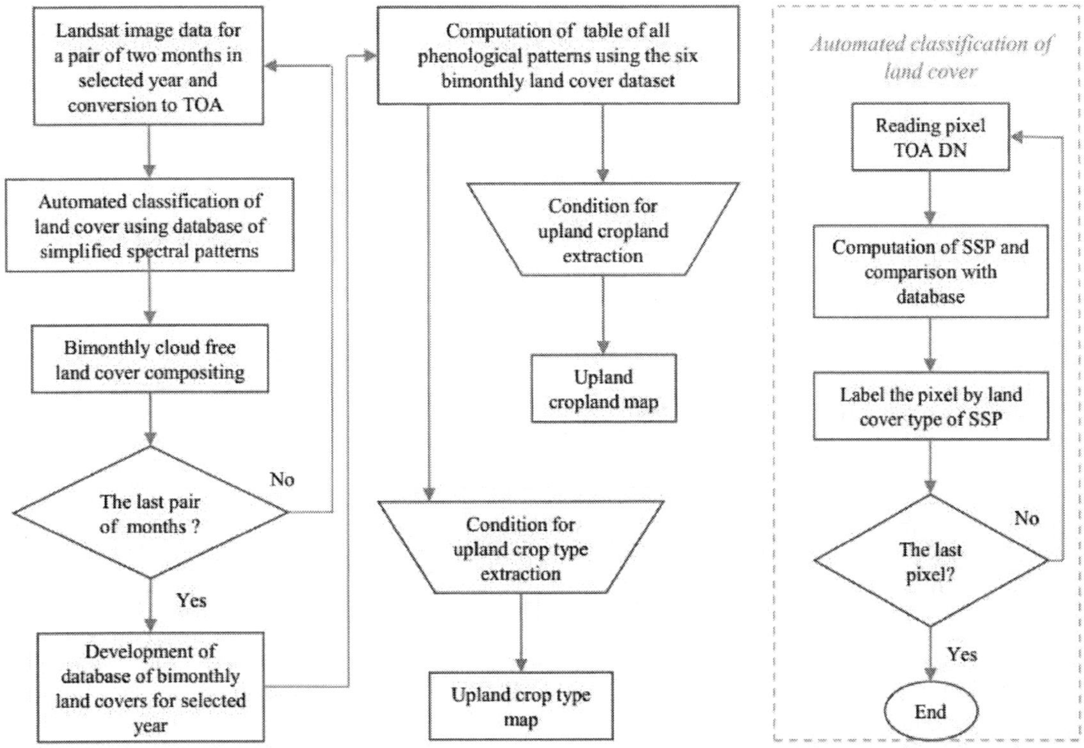

Figure 3. Block diagram of algorithm for cropland and crop type mapping with multitemporal Landsat image data. Procedure for automated classification of land cover is shown in green box in the right part.

4. RESULT AND DISCUSSION

We used 50 Landsat scenes with path/row number 128/45 (Table 1) for cropland and crop type mapping. Basically, we prefer Landsat 8 scenes from 2017, 2016 and 2015 for mapping of year 2017. However, due to high mountains in the study area which are very often covered by clouds around the year, we need to use older Landsat images taken from previous years to compensate clouds.

To support our research, we developed a software package composed of several modules as SSPC: SSP collector, ACLC: automated classification of land cover, CIC: classified image combination and PIM: phenology inspector and modelling. SSPC and PIM run in Windows environment with graphical interface allowing interaction between operator and the software. The other two modules ACLC and CIC work in command line mode offering batch processing of large amount of Landsat image data. Development of SSP database is time consuming but the use is multiple and the database needs update only when we make analysis in new geographical regions. Phenology inspection and modelling requires local knowledge and auxiliary materials to support decision of cropland and crop types classification. All modules were coded in C++ programming language compatible with all Windows operating systems. Automated classification of land cover was developed and tuned to achieve highest computing speed. Because classification of one Landsat scene takes less than 30 seconds with standard laptop computer so we can complete classification of 50 Landsat scenes in relatively short time.

The selected Landsat scenes were automatically classified by using SSP to generate land cover with 8 classes: water, wetland, close forest, open forest, brown vegetation, sparse vegetation, dry bare land and moist bare land. Cloud and cloud shadow mask provided in Quality assessment band (BQA band) of Landsat Collection 1 Level-1dataset were used to exclude bad pixels and replace them by good pixels from land cover of other date. Good and bad pixel are defined from cloud and cloud shadow contamination point of view. When combining good pixels, higher priority is given to those that are more closed to base observation date (2017) and to bare land. Reason that we assign bare land higher priority than vegetation is to catch annual growing cycle of crop. Six cloud free land covers are presented in Figure 4. By visual inspection of those land covers we can detect comprehensively seasonal changes of land covers around the year. These changes are basics for definition of phenological patterns of all crop types.

By statistical analysis, we found that there are 648,653 phenological patterns with number of pixels per pattern ranging from 1,714,258 to one. Naturally, the patterns with fewer pixels are caused by many reasons like data noise, cloud and cloud shadow contamination, misclassification and unidentifiable land cover types produced by mixing of different crop types. Therefore, we take into consideration in mapping only phenological patterns that have reasonably high number of pixels.

To map upland cropland, we set condition that during the year the pixel is never classified to water or wetland. At least at one time the pixel is classified as bare land and at one time as close or open forest. The idea behind requirement that at one time the pixel has to be bare land denotes harvesting time what is the main characteristics for annual crops. The existence of bare land and vegetation (forest) in the phenological pattern simply indicates growing cycle of annual crop. Figure 5 shows map of upland cropland. Upland cropland is coloured in red and displayed over background made by Landsat 8 colour composite RGB: 654. The background is blurred to highlight upland cropland. Distribution of cropland as seen in Fig. 5 is spanned on west site of Hoang Lien Son range where topography and climate condition is favourable for crop cultivation. Cropland spreads over slope land of low mountains with elevation from 200m to 1000m and follow relief of terrain. The area inside the selected window (Fig. 5) where upland crop is the most intensive experienced in recent years many natural hazards as flush flood and landslides. Upland cropland mapping, therefore, is important for environmental protection and natural hazards mitigation.

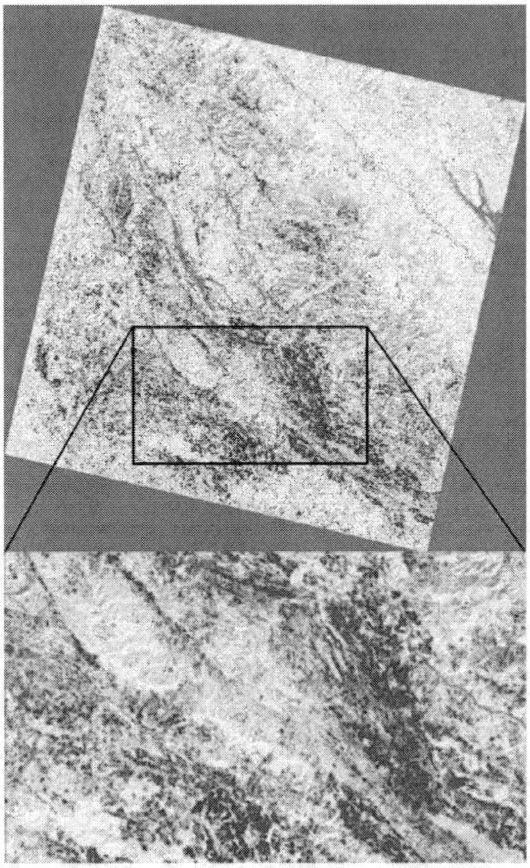

Figure 5. Cropland map of the study area and window for detail visualization. Cropland is coloured in red and displayed over blurred background made by Landsat 8 colour composite RGB: 654

Because number of phenological patterns in the study area is very big so we need to reclassify them to group of similar patterns to single class. By grouping close and open forest to new class vegetation; brown, sparse vegetation, dry bare land to bare land we can reduce substantially number of phenological patterns that could be considered as different crop types. Each potential crop is named accordingly to growing cycle with harvesting and peak growing season. For example, crop with growing (vegetative) season from May to December while from January to May the field is under harvesting, propagation, or rooting is named as BBVVVV. By this naming system we can name all phenological patterns and count their number of pixels. Due to mixture of different crop types, intensive farming practice and small size of fields, many phenological patterns are not identifiable to any crop type. We select only patterns with highest number of pixels for crop type mapping. In this study, we select 14 phenological patterns (Table 2.) with highest number of pixels ranging from 1,612,698 (145,143 ha) to 37,479 (3,373 ha) for crop identification. In this table we present list of classes that are named according to its phenological patterns and colours used for class visualization. Class BBVVVV has highest number of pixels of 1,612,698 which is equivalent to areas of 145,143 ha. The class BBVVVV can be explained that in months from January to May (dry season) fields are under either harvesting, land preparation or propagation. From May to December, crops grow and the fields become covered by green vegetation. When compared to agricultural calendar of the study site we could interpret this class as maize. The class BBBVVB stands for phenology that bare land is field status of the half of the year and in November and December. Vegetation dominates from July to October. This phenology could be assigned to upland rice. The class BVVVVB has phenology where vegetation dominates in months from March to November. Harvesting is either on the beginning or on the end of the year. This phenology is typical for sugarcane cultivation.

Class Name	Class Code	Number of Pixels per class	Colour
BBVVVV	5	1612698	
VBVVVV	6	1460601	
BVVVVV	7	846403	
VVBVVV	8	500020	
BBBVVV	9	475380	
BBBVVB	10	459834	
VBBVVV	11	226421	
BVVVVB	12	127938	
VBVVVB	13	95366	
BBBVBB	14	84296	
BBVVBB	15	66791	
BBVBVB	16	55994	
BVBVVV	17	53713	
BBBBVB	18	37479	

Table 2. Fourteen crop types with highest number of pixels selected for crop type mapping

Figure 6 shows crop type map and enlarged window showing details of distribution of different crops. Colours used for presentation of classes are given in Table 2. In this figure, crop types with similar phenological patterns are visualized by similar colour. Similar colour formulates small and fragmented patches showing spatial spreading of crops. We can notice quite clear spatial patterns of distribution of crops, which might reflect cropping practice. Assessment of spatial distribution of crops reveals that crop that has similar phenological patterns are

This contribution has been peer-reviewed.
https://doi.org/10.5194/isprs-archives-XLII-3-W7-11-2019 | © Authors 2019. CC BY 4.0 License.

spreading with patterns showing local farming practices. Class BBVVVV is major crop in the study area, coloured in red in Fig. 6, is interpreted as maize. Maize cultivation, however like many other crops, does not have fixed date of propagation and consequently fixed date of harvesting. There are early and late maize and time tolerance of propagation could be up to four months. This practice and trend in intercropping of maize and cassava recently result complexness in using phenological patterns for crop identification. However, we expect that phenological patterns can provide us spatial information of major crops in study area. Our study area is remote, mountainous and undeveloped region with limited access. Moreover, very high spatial resolution satellite imagery is also not available in Google Earth so validation of our cropland and crop type mapping is limited. Nevertheless, our approach is very promising and the proposed algorithm needs to be continued and tuned in future.

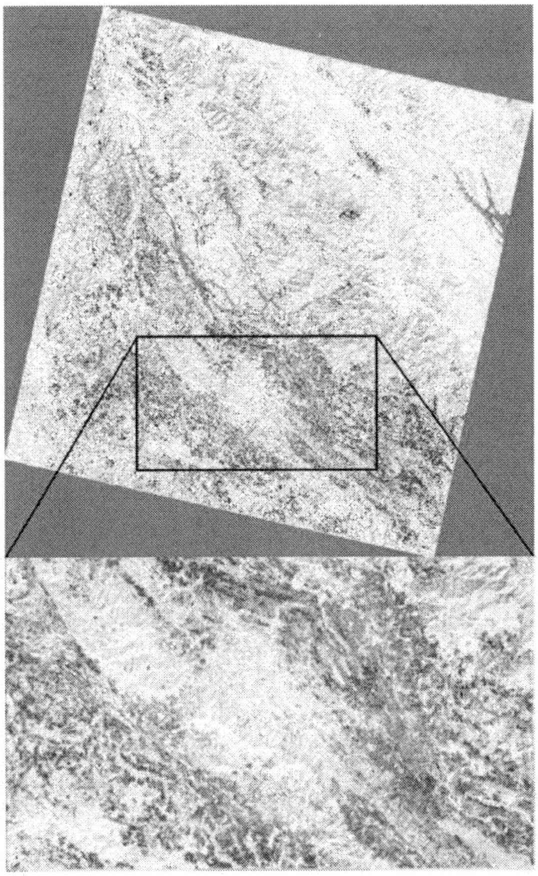

Figure 6. Map of crop types and window for detail visualization. There are 14 crop types dyed by colours in Table 2. Blurred background image is made by Landsat 8 colour composite RGB: 654

5. CONCLUSION

Cropland and crop type mapping is one of major application of multitemporal optical remote sensing. The NDVI time series of coarse remote sensor like MODIS are utilized as main data source for cropland and crop type mapping. However, for better land use management, the use of higher spatial resolution remote sensing data such as Landsat is getting in demands. In this study, we present development of new algorithm for cropland and crop type mapping using multitemporal Landsat images. Our proposed algorithm can do analysis with image data of all Landsat sensors including Landsat 4/5, Landsat 7 and Landsat 8. Image data collected during one year is grouped to six clusters representing bimonthly observation. Images within each cluster are automatically classified to numbers of land cover map that are subsequently combined together to exclude clouds and cloud shadows. The six bimonthly cloud free land covers are used further for development of phenological pattern database. During the development of database, statistical computation is applied to identify all phenological patterns. These phenological patterns are main input for cropland and crop type mapping. Due to inevitability of cloud cover affecting quality of image data of the selected year, we need to use Landsat image data from years back to compensate cloud and cloud shadow and that could influence reliability of the final results, especially for mountainous region with high probability of cloud around the year. By using our proposed algorithm, we success to map cropland and crop type distribution in the study area covered by Landsat scene with path/row number 128/45. The results we achieved are created by application of logical models with phenological conditions for cropland and crop types that dominantly spread in the study area. Different phenology based land cover types such as bare land or cover of distinct vegetative status is assumed to represent different crop types. As result, we identified 14 major crop types with area ranging from 145,143 ha to 3,373 ha. Because the study area is located in mountainous area without any map of crop distribution so we could not carry out accuracy assessment. However, visual interpretation of our map overlaid on false colour composite of Landsat images reveals that cropland distribution is quite reliable and acceptable. Due to lack of crop type distribution map we could not assess quality of crop type mapping. However, spatial analysis of distribution of crop fields demonstrates logical accordance of crop type placements with terrain characteristics and cropping practices. Finally, the study pointed out that phenology characterized by six bimonthly land covers is good enough to identify cropland distribution and crop type classification. To achieve better results, apparently we need image data with higher temporal resolution. Application of multitemporal SAR data could be a way to overcome this obstacle.

ACKNOWLEDGEMENTS

The authors would like to acknowledge support from the Vietnam National Foundation for Science and Technology Development (NAFOSTED) and the Vietnam Academy of Science and Technology for this research.

REFERENCES

Brenning, A., Kaden, K., Itzerott, S., 2006. Comparing classifiers for crop identification based on multitemporal landsat TM/ETM data. Proc. 2nd Work. EARSeL SIG L. Use L. Cover 64–71.

Chen, Y., Lu, D., Moran, E., Batistella, M., Dutra, L.V., Sanches, I.D., da Silva, R.F.B., Huang, J., Luiz, A.J.B., de Oliveira, M.A.F., 2018. Mapping croplands, cropping patterns, and crop types using MODIS time-series data. Int. J. Appl. Earth Obs. Geoinf. 69, 133–147. https://doi.org/10.1016/j.jag.2018.03.005

Csillik, O., Belgiu, M., 2017. Cropland mapping from Sentinel-2 time series data using object-based image analysis. Agil. 2017 2016.

Deng, C., Wu, C., 2013. ISPRS Journal of Photogrammetry and Remote Sensing The use of single-date MODIS imagery for estimating large-scale urban impervious surface fraction with spectral mixture analysis and machine learning techniques. ISPRS J. Photogramm. Remote Sens. 86, 100–110. https://doi.org/10.1016/j.isprsjprs.2013.09.010

Dong, J., Xiao, X., Menarguez, M.A., Zhang, G., Qin, Y., Thau, D., Biradar, C., Moore, B., 2016. Mapping paddy rice planting area in northeastern Asia with Landsat 8 images, phenology-based algorithm and Google Earth Engine. Remote Sens. Environ. 185, 142–154. https://doi.org/10.1016/j.rse.2016.02.016

Duong, N.D., 2018. Decomposition of Landsat 8 OLI Images by Simplified Spectral Patterns for Land Cover Mapping. 2018 10th IAPR Work. Pattern Recognit. Remote Sens. 1–13.

Duong, N.D., 2016. Automated classification of Land cover using Landsat 8 OLI Surface Reflectance product and spectral pattern analysis concept - Case study in Hanoi, Vietnam. Int. Arch. Photogramm. Remote Sens. Spat. Inf. Sci. - ISPRS Arch. 41, 987–991. https://doi.org/10.5194/isprsarchives-XLI-B8-987-2016

Duong, N.D., Anh, L.V., Thu, H.L., Anh, N.K., 2014. Spectral signatures in landsat 8 oli image and their interpretation for land cover study, in: 35th Asian Conference on Remote Sensing 2014, ACRS 2014: Sensing for Reintegration of Societies.

Duong, N.D., Hang, L.M., Tuan, T.A., Ouyang, Z., 2017. Development of a spectral-pattern-analysis-based method for automated water body extraction using Landsat image data: A case study in central Vietnam and southern Laos. Limnol. Oceanogr. Methods 15, 945–959. https://doi.org/10.1002/lom3.10215

Foerster, S., Kaden, K., Foerster, M., Itzerott, S., 2012. Crop type mapping using spectral-temporal profiles and phenological information. Comput. Electron. Agric. 89, 30–40. https://doi.org/10.1016/j.compag.2012.07.015

Shao, Y., Lunetta, R.S., Ediriwickrema, J., Liames, J., 2009. Mapping Cropland and Major Crop Types across the Great Lakes Basin using MODIS-NDVI Data. Photogramm. Eng. Remote Sensing 76, 73–84. https://doi.org/10.14358/PERS.76.1.73

Simonetti, D., Simonetti, E., Szantoi, Z., Lupi, A., Eva, H.D., 2015. First Results from the Phenology-Based Synthesis Classifier Using Landsat 8 Imagery. IEEE Geosci. Remote Sens. Lett. 12, 1496–1500. https://doi.org/10.1109/LGRS.2015.2409982

Simonetti, E., Simonetti, D., Preaton, D., 2014. Phenology-based land cover classification using Landsat 8 time series, European Commission Joint Research Centre - Institute for Environment and Sustainability. https://doi.org/10.2788/15561

Tatsumi, K., Yamashiki, Y., Canales Torres, M.A., Taipe, C.L.R., 2015. Crop classification of upland fields using Random forest of time-series Landsat 7 ETM+ data. Comput. Electron. Agric. 115, 171–179. https://doi.org/10.1016/j.compag.2015.05.001

Victoria, D. de C., da Paz, A.R., Coutinho, A.C., Kastens, J., Brown, J.C., 2012. Cropland area estimates using Modis NDVI time series in the state of Mato Grosso, Brazil. Pesqui. Agropecu. Bras. 47, 1270–1278. https://doi.org/10.1590/S0100-204X2012000900012

Waldner, F., Hansen, M.C., Potapov, P. V., Löw, F., Newby, T., Ferreira, S., Defourny, P., 2017. National-scale cropland mapping based on spectral-temporal features and outdated land cover information. PLoS One 12, 1–24. https://doi.org/10.1371/journal.pone.0181911

Wezel, A., Luibrand, A., Friederichsen, J.R., Thanh, L.Q., 2006. Soil fertility , crop production and temporal changes of resource use on steep slopes in upland northwest Vietnam 1–13.

Xiong, J., Thenkabail, P.S., Gumma, M.K., Teluguntla, P., Poehnelt, J., Congalton, R.G., Yadav, K., Thau, D., 2017. Automated cropland mapping of continental Africa using Google Earth Engine cloud computing. ISPRS J. Photogramm. Remote Sens. 126, 225–244. https://doi.org/10.1016/j.isprsjprs.2017.01.019

Zhao, X., Wang, X., Cao, G., Chen, K., Tang, W., Zhang, Z., 2017. Crop Identification by Using Seasonal Parameters Extracted from Time Series Landsat Images in a Mountainous Agricultural County of Eastern Qinghai Province, China. J. Agric. Sci. 9, 116. https://doi.org/10.5539/jas.v9n4p116

Zhong, L., Hu, L., Yu, L., Gong, P., Biging, G.S., 2016. Automated mapping of soybean and corn using phenology. ISPRS J. Photogramm. Remote Sens. 119, 151–164. https://doi.org/10.1016/j.isprsjprs.2016.05.014

This contribution has been peer-reviewed.
https://doi.org/10.5194/isprs-archives-XLII-3-W7-11-2019

CHARACTERISTICS OF THE DEGREE OF GRADE IN GRADE-ADDED ROUGH SET FOR LAND COVER CLASSIFICATION

Y. Ishii [1], K. Iwao [2], T. Kinoshita [3,*]

[1] Graduate School of Agriculture, Ibaraki University, 300-0393, Ibaraki, Japan - 18am401h@vc.ibaraki.ac.jp
[2] Geological Institute of Geology and Geoinformation Geological Survey of Japan, Advanced Industrial Science and Technology, 305-8567, Ibaraki, Japan - iwao.koki@aist.go.jp
[3] College of Agriculture, Ibaraki University, 300-0393, Ibaraki, Japan - tsuguki.kinoshita.00@vc.ibaraki.ac.jp

Commission III, WG III/10

KEY WORDS: Grade-added rough set, Maximum likelihood classifier, Support vector machine, Membership function, Land cover classification, Multispec

ABSTRACT:

This paper aims to clarify the meaning of the membership which is produced as by-products of land cover classification by Grade-added rough set (GRS). A new land cover classification method by using GRS was developed. The classification scheme of GRS which calculates membership (degree of grade) for each class is similar to those of MLC and SVM. But there are two things that are not clear. One is a meaning of the membership of GRS and the other is a reason why the larger membership in GRS employed works well. In this study, aerial images were used to visualize the relation of membership between GRS and existing classifiers, MLC and SVM. Furthermore, a model experiment in two-dimensional feature space was conducted. From these experiments, it was found that the meaning of degree of grade is a distance from a nearest training data of other class. That is, the meaning of membership of GRS is similar to that of SVM, because SVM also calculates a distance from boundary line which is determined by support vectors, while the meaning of membership of MLC is a distance from a centroid of own class. Also it was found that what the distance from the closest other class is given as the degree of grade implies that the higher the grade, the higher the certainty. In this research we could clarify some of the features of land cover classification using GRS.

1. INTRODUCTION

A new land cover classification method called Grade-added rough set (GRS) was developed (Ishii *et al.*, 2018). The rough set theory which is the basis of this approach was proposed by Pawlak (Pawlak, 1982).

In the process of land cover classification, the three methods, GRS, MLC and SVM have a similarity that it outputs a map of membership for each class and then classifies each pixel into an appropriate class referring to the membership. In GRS, the grade degree of each class is obtained for each pixel and classified into the class of the maximum grade degree. Pixels are allocated to their most likely class of membership in MLC (Foody *et al.*, 1992). In SVM, pixels are classified a class which the decision function takes maximum value when separating the class from the remaining class. In this way, these three classification methods once calculate the membership of each class and compare the membership with other classes. However, it is not clear why classifying to the class whose grade degree is maximum. Therefore, by examining the relation between GRS, MLC, and SVM membership value, we clarify what grade degree in GRS means in land cover classification.

In this research, we clarify the relation of membership between GRS and existing method, MLC and SVM using actual aerial photographs, then reveal the relation in two-dimensional modelled feature space, comparing the characteristics of membership values of GRS, MLC and SVM. This paper aims to clarify the meaning of the degree of grade of GRS in the land cover classification through these experiments.

2. MEMBERSHIP FUNCTIONS

We introduce the method of calculating the membership of the three classifiers used in this study.

2.1 GRS

GRS is a method developed based on the rough set theory. It is known that the classification will be successful to allocate a pixel into a class which is the highest degree of grade (Ishii *et al.*, 2018). The degree of grade is obtained as follows (Mori, et al., 2004). A decision matrix is created for training data belonging to class k and not belonging to class k. M_{ij}^k is an element of decision matrix for class k.

$$M_{ij}^k = \{(a, \rho(\boldsymbol{x}_i, a), g_{ij}) | \rho(\boldsymbol{x}_i, a) \neq \rho(\boldsymbol{x}_j, a)\}, \quad i \in K_k^+, j \in K_k^- \tag{1}$$

where, $\boldsymbol{x}$ is training data, a is an attribute, ρ is an attribute value, g_{ij} is a grade degree which calculated by $g_{ij} = \rho(\boldsymbol{x}_i, a) - \rho(\boldsymbol{x}_j, a)$, $K_k^+ = \{i | \boldsymbol{x}_i \in C_*(D_k)\}$, and $K_k^- = \{i | \boldsymbol{x}_i \notin D_k\}$. A threshold is defined as α. If $|g_{ij}| \geq \alpha$, regard the sample as discernible from the other class. Then, calculate the disjunction about each attribute,

$$\mathcal{L}(M_{ij}^k) = (\rho(\boldsymbol{x}_i, a_1) \vee \rho(\boldsymbol{x}_i, a_2) \vee \cdots \vee (\boldsymbol{x}_i, a_m), arg\,max(|g_{ij}|)) \tag{2}$$

* Corresponding author

https://doi.org/10.5194/isprs-archives-XLII-3-W7-19-2019

It is necessary to be discernible with all training data of other class, so calculate the conjunction. Furthermore, calculate the disjunction in order to need to be distinguished by at least one training data of class k. $\boldsymbol{x}$ is redefined as each pixel of image.

$$D_k(\boldsymbol{x}) = \bigvee_{i \in K_k^+} \bigwedge_{j \in K_k^-} \mathcal{L}\,(M_{ij}^k) \quad (3)$$

where calculate arg $min(|g_{ij}|)$ in conjunction and arg$max(|g_{ij}|)$ in disjunction. For each class, calculate eq. (3), and each pixel allocates the class which maximize eq. (3). That is,

$$f(\boldsymbol{x}) = \arg max_{i=1,\dots,n}\, D_i(\boldsymbol{x}) \quad (4)$$

If a pixel is not allocated any class or there are more two class of max membership, the pixel is allocated 0.

2.2 MLC

MLC is a method of finding likelihood and classifying pixels into a class that maximizes its likelihood, but actually it is a problem of classifying it into a class that minimizes the following formula derived on the basis of likelihood (Japan Remote Sensing Society, 2011).

$$D_k(\boldsymbol{x}) = ln|\boldsymbol{\Sigma}_k| + (\boldsymbol{x} - \boldsymbol{m}_k)^T \boldsymbol{\Sigma}_k^{-1}(\boldsymbol{x} - \boldsymbol{m}_k) \quad (5)$$

where, each pixel is $\boldsymbol{x}$, a mean vector of class k is $\boldsymbol{m}_k$, variance-covariance matrix of class k is $\boldsymbol{\Sigma}_k$.

$$f(\boldsymbol{x}) = \arg min_{i=1,\dots,n}\, D_i(\boldsymbol{x}) \quad (6)$$

2.3 SVM

Let the i th decision function separating class i from the remaining class be given by

$$D_k(\boldsymbol{x}) = \boldsymbol{w}_k^T \phi(\boldsymbol{x}) + b_k \quad (7)$$

where, $\boldsymbol{x}$ is each pixel, $\phi(\boldsymbol{x})$ is a function of mapping input space to l dimensional feature space, $\boldsymbol{w}_k$ is l-dimensional coefficient vector and b_k is bias (Abe, 2011). For the 2-class problem, the sgn function of Eq. (7) becomes the decision function, but in the case of multi-class classification, the following function is the decision function.

$$f(\boldsymbol{x}) = \arg max_{i=1,\dots,n}\, D_i(\boldsymbol{x}) \quad (8)$$

3. METHODS

3.1 Data sets

We used A Moderate Dimension Example published by A Freeware Multispectral Image Data Analysis System of Purdue University (Landgrebe and Biehl, 1994). This is a data set of 12 bands taken from the aircraft at a farmland in Indiana in June 1966. The size of the image is 949 × 220 pixels, of which 70594 points are certain data whose classes are known.

Based on the experimental results using this data set by Purdue University (Landgrebe, 1997), we used 4 bands of 12 bands, Band 1, 6, 9 and 10. Based on the central limit theorem, about 70,000 points of data were divided into 31. 30 training data sets and a validation data set were created. Table. 1 shows class name, the number of pixels and the area proportion of 70594 points for each class. The difference between the minimum number of class and the maximum number of class is about 20 times. Since 31 divisions were conducted randomly, there are some differences in the area proportion of each class by the training data set, but it is almost the same as the data set.

Table 1. Multispec test data set

No.	Class	Pixels	Area proportion (%)
1	Alfalfa	3375	4.8
2	Bare Soil	1230	1.7
3	Corn	10625	15.1
4	Oats	5781	8.2
5	Red Clover	12147	17.2
6	Rye	2385	3.4
7	Soybeans	25133	35.6
8	Wheat 1	7827	11.1
9	Wheat 2	2091	3.0

3.2 Experimental design

In this study, land cover classifications using Multispec data set were conducted by three methods, GRS, MLC, and SVM. Then, the maps of membership are outputted for each class. From the membership map, 70549 points whose correct classes were known were extracted. When classification was conducted, the data set of the original 8-bit Multispec was normalized a mean to 0 and variance to 1, and then it multiplied the number by 1000. Assume a threshold $\alpha = 1$ in GRS. Radial basis function (RBF) was used for the SVM kernel. In this paper, hyperparameters of SVM were set to $\sigma = 1.0$ and regularization parameter $C = 100.0$. This classification was repeated for 30 data sets. Overall accuracy of maps was evaluated, because it is better to know how accurate each classifier is for examining the correlation of membership.

Furthermore, experiments were conducted to investigate the differences of three classifiers using a two-dimensional feature space. We assumed two classes and examined how each classifier draws a class boundary. Training data of each class was set to five (1st row in Fig. 3). It is known that MLC is a method considering variance of each class, but it is not clear GRS and SVM can consider the effect of variance of each class. So, experiments were carried out with three variance cases (1st row in Fig. 3) to investigate whether GRS and SVM have such effect.

4. RESULTS AND DISSCUSION

Land cover classifications were conducted by GRS, MLC and SVM using Multispec data set. Fig. 1 shows overall accuracy with 30 training data sets. SVM has the highest average accuracy, but the variance is large. On the other hand, MLC has the smallest variance.

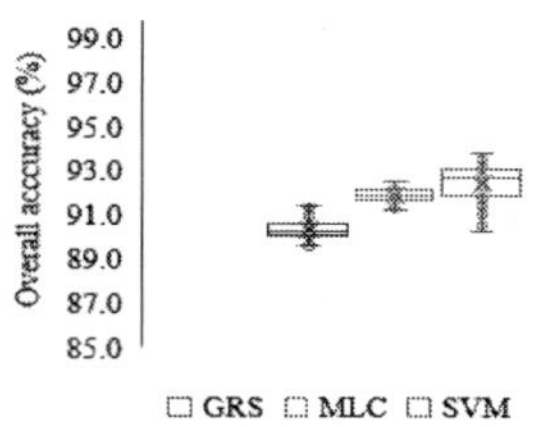

Figure 1. Overall accuracy

https://doi.org/10.5194/isprs-archives-XLII-3-W7-19-2019 | © Authors 2019. CC BY 4.0 License.

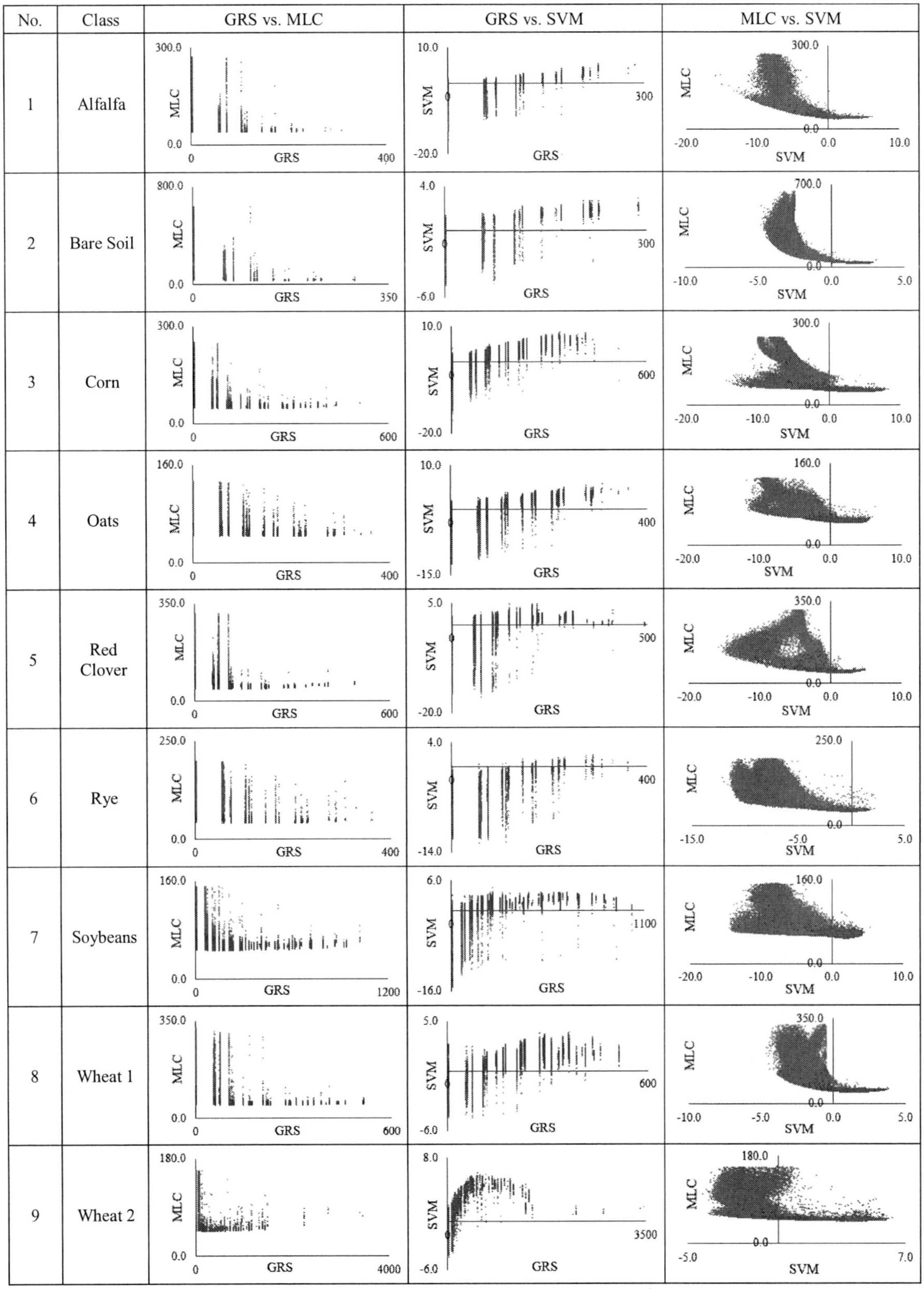

Figure 2. Scatter plots of membership between two classifiers

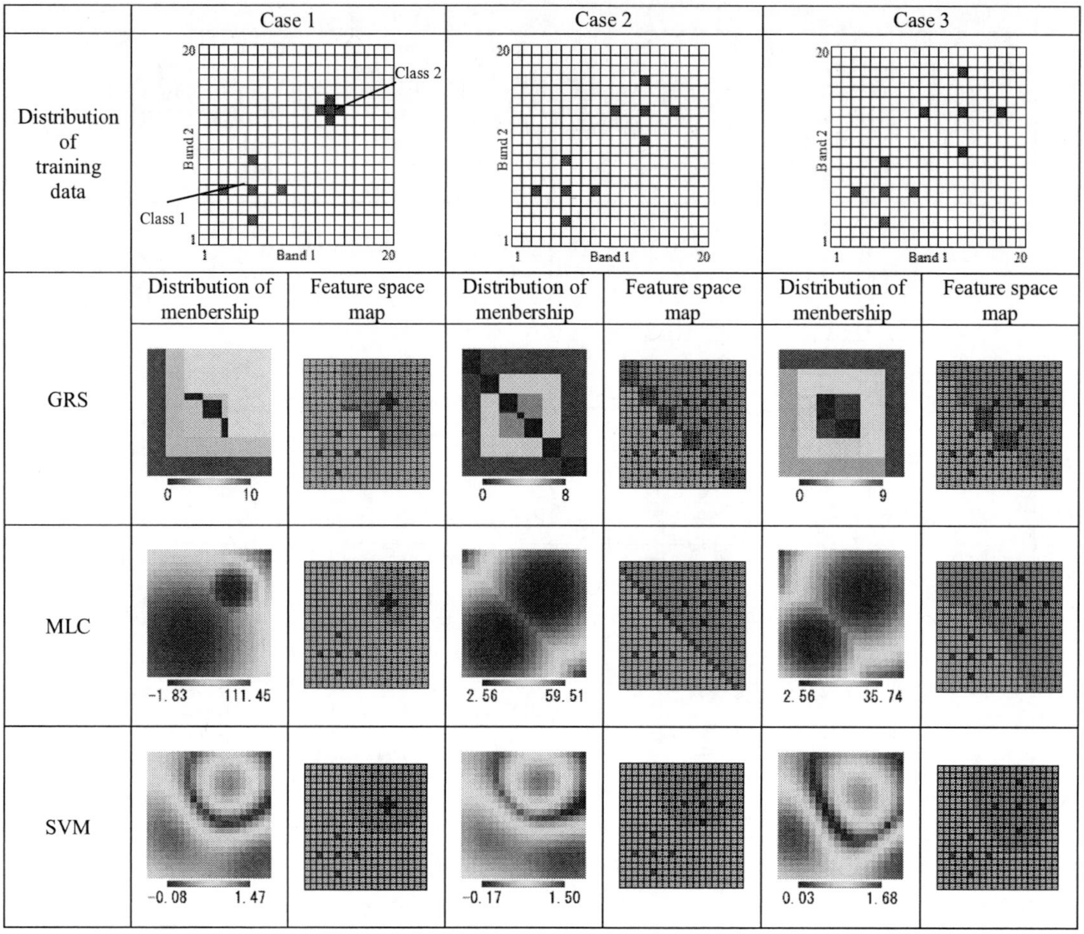

Figure 3. Distribution of training data, membership and classification map in feature space in three cases

Then, using Multispec image, a trend of membership was examined. Each classifier outputted nine class membership maps for each 30 training data sets. In addition, the correlation of membership of GRS and MLC, GRS and SVM, MLC and SVM was examined for each class. The number of points used for examining the correlation were 70594 points where the correct class was known. As a result, similar results were shown from different training data sets, so only a result of one training set is represented in Fig. 2. The graphs in Fig. 2 are scatter plots of membership of two classifiers. The horizontal axis indicates a membership of one classifier, and the vertical axis indicates that of the other. Although the membership of GRS is integer value, that of MLC and SVM is continuous value. The red points in each graph are the membership which is actually correct answer, and grey points are the membership not chosen that class. There are about 70,000 points in each graph, but some points overlap.

Focusing on the distribution of correct classes (red points) in GRS and MLC, the membership of MLC is mostly low, but the membership of GRS takes various values. On the other hand, paying attention to the distribution of grey points, the membership of MLC exponentially increases as the membership of GRS decrease.

Many of GRS and SVM have slightly positive correlation. However, as the membership of GRS gets lager, the membership of SVM become lager, but there are also cases where membership of SVM is slightly lower when GRS membership is lager.

In MLC and SVM, they have a common point that there is a distribution of the correct class (red points) in the lower right to a distribution of all points. In other words, when the membership of SVM is small, MLC takes wide range of membership, but when the membership of SVM is large corresponds to only when the membership of MLC is small. This means that the membership of MLC and SVM almost agrees with reasoning for correct class, but the membership of other classes shows different prediction depending on the deference of classification scheme.

Next, Fig. 3 shows the results of experiments using a two-dimensional feature space. The horizontal axis represents band 1, and the vertical axis represents band 2. A range of value is from 1 to 20. As shown in the first row of Fig. 3, three training data sets including different variance class are considered. It was assumed that blue training data is class 1 and red training data is class 2. Case 1 has smaller variance of class 2 than class 1. Case 2 has the same variance of class 1 and class 2. Case 3 has larger variance of class 2 than class 1. In all cases, the centroids of the classes are fixed. The rows from second to fifth in Fig. 3 represent the distribution of membership (left side) and the map of the classification result in feature space (right side).

As a result, the degree of grade becomes smaller towards the center of the feature space in GRS, and when the degree of grade is the same, the point is unclassified. Unclassified points are stored 0. The degree of grade of both classes in the vicinity of the boundary line is a little different, and a class having a higher degree of grade is selected. Observing the degree of grade of each cell inferred from the training data revealed that the degree of grade stored in each cell indicates the distance from the closest training of the opponent class in the feature space. Therefore, it was revealed that to calculate the degree of grade is to find the closest training data of the opponent class, and the degree of grade shows how far from the closest training data of the opponent class. Extracting rules by disjunction of training data of the same class means to adopt the point with the larger degree of grade. This indicates that the larger grade of degree, the larger distance to the closest training data of opponent class. Furthermore, it can be interpreted that more reliable information is adopted because this is more reliable training data as the distance to the opponent is greater. That is, the greater the degree of grade, the higher the certainty of being their own class. In classification by GRS, a map of membership for each class is created, and among them each point classifies into a class with the highest grade of degree. It was confirmed that the reason for this is that the degree of grade can be regarded as equivalent to the degree of certainty. Also, the logic that unclassification occurs at the boundary because the certainty of both classes is of equal degree is reasonable. In addition, it can be said that GRS can take into account the problem that whether variance can be considered. In fact, the concept of variance does not appear in the GRS. However, comparing when the variance of class 2 is large and when it is small, the degree of grade stored in class 1 is different. For example, when the centroids are fixed, the larger the variance of class 2 is, the closer the distance is to the side closer to class 1 than class2's centroid, and the further away from class 1 than class2's centroid, the further the distance becomes. This indicates what an operation of taking a disjunction for rules extracted from each training data of the same class is easier to adopt lager grade of degree by a relatively large variance class and smaller grade of degree by a relatively small variance class. As a result, it can be seen from the second row in Fig. 3 that the class with large variance has a scheme that dominates the relatively large region in the feature space as compared with the small class.

Next, in the case of MLC, it was confirmed that the membership increases as the distance from the centroid obtained from the training data, and the class boundary is drawn where the magnitude of the membership changes between classes. MLC is a method that takes variance and inclination into consideration from the theory. From the results in Fig. 3, it is also found that the change with the distance from the centroid is larger for classes with smaller variance. And grey cellls indicates unclassified which occurs by same membership in Case 2. In addition, MLC is fundamentally similar to the nearest neighbour method, and it is a different property from GRS in that it calculates how close the distance from the centroid of opponent class is.

Fig. 4 shows a graph of the distribution of membership of GRS and MLC on class 1 of Case 1 on a line AB in the feature space. Fig. 4 (a) is a position of line AB in feature space, and Fig.4 (b) is membership of GRS and SVM on a line AB. Normalization that the average is 0 and variance is 1 for each membership, was performed, and then, took minimum value from all membership to make them positive numbers. From Fig. 4, we can explain the relation between GRS and MLC in Fig. 2. When the membership of the MLC takes a relatively low value, it takes a variety of membership in the GRS in Fig. 2. That coincides with the graph to the left side in Fig. 4 (b). In addition, when the membership of GRS takes a relatively small value, MLC has a wide range of values. That coincides with the graph to the right side Fig. 4 (b).

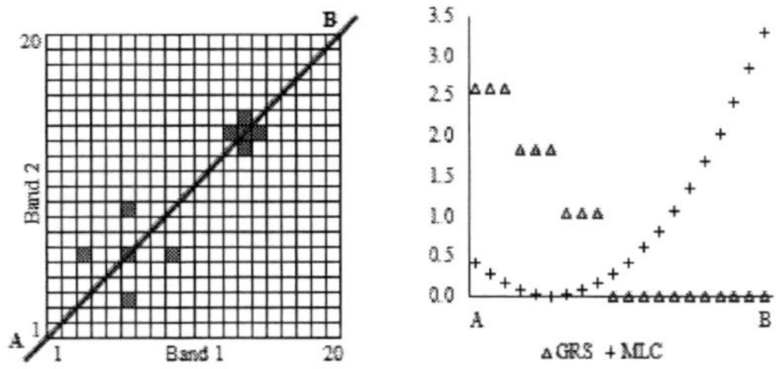

(a) Position of line AB in feature space (b) Membership of GRS and MLC on a line AB

Figure 4. Comparison of class 1 membership between GRS and MLC in Case 1

In the case of SVM, it was confirmed that the boundary line is first obtained from the support vector, then the distance from it is given as membership, and as the result of adopting a class with a large distance, the boundary is finally drawn by change points of the magnitude of membership between classes. In SVM, it was not clear from this experimental result whether the effect of class variance was taken into consideration.

Fig. 5 (b) is a graph of comparing the distribution of Class 1 membership on a line AB in the feature space between GRS and SVM in Case 1. The trend is similar between GRS and SVM. Fig. 5 (b) explains to the tendency of distribution between GRS and SVM in Fig. 2.

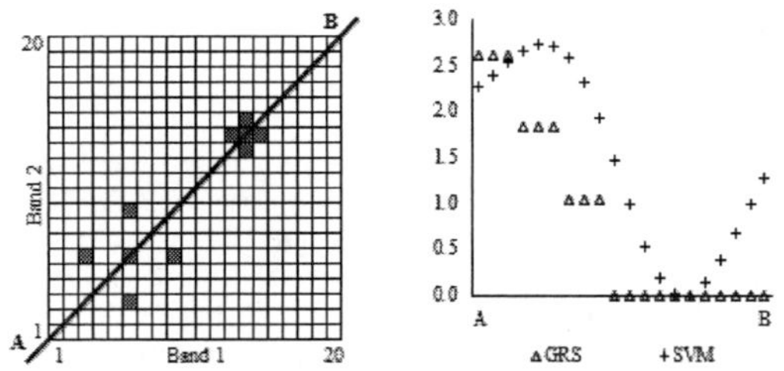

(a) Position of line AB in feature space (b) Membership of GRS and SVM on a line AB

Figure 5. Comparison of class 1 membership between GRS and SVM in Case 1

The degree of grade in GRS is the distance from the closest training data of the other class, and the membership is the distance from the boundary obtained by the support vector which is the closest distance to the other class. That is, similar in that the distance is obtained from the relation with other class. Meanwhile, the membership of MLC is the distance from the centroid of its own class. From this point it differs greatly from GRS and SVM. (Foody and Mathur, 2006) and (Brown, Lewis and Gunn, 2000) have already used the membership value of SVM to solve the mixed pixel problem. Therefore, what the way of finding membership in GRS is similar to that of SVM means that GRS has the possibility of applying to mixed pixel estimation. GRS has advantages which does not need the adjustment of hyperparameter like SVM, so we plan to consider GRS application to mixed pixel estimation in the future.

This contribution has been peer-reviewed.
https://doi.org/10.5194/isprs-archives-XLII-3-W7-19-2019 | © Authors 2019. CC BY 4.0 License.

5. CONCLUSIONS

GRS, MLC, SVM have a common point in classifying based on the membership of each class for land cover classification. In this paper, we discussed the significance of membership of each class generated by land cover classification using GRS, MLC or SVM. First, using the Multispec data set published by Purdue University, land cover classification is performed with these three classifiers. Then, membership maps of each class were output, and the distributions of membership of GRS vs. MLC, GRS vs. SVM and MLC vs. SVM are examined. Although these results can not be said to be completely correlated, they showed that they are any relations to each other. In order to investigate about these relations, a model experiment was conducted in a two-dimensional feature space. As a result of classification of 2 classes with 5 training data per class, the characteristics of drawing boundary in the feature space of each classifier was clarified. In GRS, the distance between its own class and the closest point of other class is given as the degree of grade. In addition, comparing the degree of grade of the two classes and adopting the larger class means to adopt a class with a higher certainty class. In MLC, it was confirmed that a boundary was drawn where the distance as the magnitude of the value adding the effect of variance and inclination from the centroid of its own class changes between two classes. In SVM, it was confirmed that the distance from the boundary obtained from the support vector is given as a membership value, and the place where the distance from the line changes is the boundary in the feature space. It is concluded that the way of classification of GRS is close to SVM in that it focuses on the distance to the other class. SVM is already used for estimating mixed pixels. So there is a possibility to use GRS as a more practical method for estimating mixed pixels.

REFERENCES

Abe, S., 2011. Introduction to support vector machine for pattern recognition. *Morikita publishing*, 224pp. (in Japanese).

Brown, M., Lewis, H. G. and Gunn, S. R., 2000. Linear Spectral Mixture Models and Support Vector Machines for Remote Sensing. *Bayrischer Rundfunk*, 38(5), pp. 2346–2360.

Foody, G. M., Campbell, N. A., Trodd, N. M. and Wood, T. F., 1992. Derivation and applications of probablistic measures of class membership from maximum-likelihood classification. In: *Photogrammetric Engineering & Remote Sensing*, 58(9), pp. 1335–1341. doi: 10.3851/IMP1936.

Foody, G. M. and Mathur, A., 2006. The use of small training sets containing mixed pixels for accurate hard image classification: Training on mixed spectral responses for classification by a SVM. In: *Remote Sensing of Environment*, 103(2), pp. 179–189. doi: 10.1016/j.rse.2006.04.001.

Ishii, Y., Bagan, H., Iwao, K. and Kinoshita, T., 2018. Comparison of land cover classification using rough set theory. In: *39th ACRS 2018 PROCEEDING*.

Japan Remote Sensing Society (Ed.), 2011. Remote Sensing Understanding from Basics. *Rikoh Tosho*, pp.242-245 (in Japanese).

Landgrebe, D., 1997. Multispectral Data Analysis: A Moderate Dimension Example©. In: *Electrical Engineering*, pp. 1–10.

Landgrebe, D. and Biehl, L., 1994. MultiSpec © A Freeware Multispectral Image Data Analysis System Documentation for MultiSpec. Purdue Research Foundation https://engineering.purdue.edu/~biehl/MultiSpec/documentation .html (9 January 2019).

Mori, N., Tanaka, H. and Inoue, K. (Eds), 2004. Rough sets and kansei. Kainundo, pp.131-181 (in Japanese).

Pawlak, Z., 1982. Rough Sets. In: *International Journal of Computer & Information Sciences*, 11(5), pp. 341–356. doi: 10.1007/978-3-7908-1776-8.

The International Archives of the Photogrammetry, Remote Sensing and Spatial Information Sciences, Volume XLII-3/W7, 2019
TC III WG III/2,10 Joint Workshop "Multidisciplinary Remote Sensing for Environmental Monitoring", 12–14 March 2019, Kyoto, Japan

A Fit-for-Purpose algorithm for Environmental Monitoring based on Maximum likelihood, Support Vector Machine and Random Forest

Ali Jamali[1]*

[1] Faculty of Surveying Engineering, Apadana Institute of Higher Education, Shiraz, Iran- ali.jamali.65@gmail.com

Commission III, WG III/10

KEYWORDS: Image classification, Earth Observation, Support Vector Machine, Random Forest, R

ABSTRACT:

Due to concerns of recent earth climate changes such as an increase of earth surface temperature and monitoring its effect on earth surface, environmental monitoring is a necessity. Environmental change monitoring in earth sciences needs land use land cover change (LULCC) modelling as a key factor to investigate impact of climate change phenomena such as droughts and floods on earth surface land cover. There are several free and commercial multi/hyper spectral data sources of Earth Observation (EO) satellites including Landsat, Sentinel and Spot. In this paper, for land use land cover modelling (LULCM), image classification of Landsat 8 using several mathematical and machine learning algorithms including Support Vector Machine (SVM), Random Forest (RF), Maximum Likelihood (ML) and a combination of SVM, ML and RF as a fit-for-purpose algorithm are implemented in R programming language and compared in terms of overall accuracy for image classification.

1. INTRODUCTION

Land cover is a fundamental factor that links and affect with many parts of the human and physical environment (Foody, 2002). The change in land cover is considered as an important factor of global change affecting ecological systems (Vitousek, 1994) with an impact on the earth that is linked with climatic change (Skole, 1994).

Land cover mapping (Grippa et al., 2018) and monitoring is one of key applications of earth observation satellites sensor data which is an important factor to asses results of climate change in the recent years. On the other hand, changes in land cover affect the climate through changes in the composition of greenhouse gasses such as carbon dioxide (Betts et al., 2007; Bonan, 2008; Bala et al., 2007).

According to Rodriguez-Galiano et al. (2012), there are several issues for large area land cover monitoring including:

1. First, complex landscapes are difficult to monitor due to sudden changes in environmental gradients (e.g. moisture, elevation and temperature) and a legacy of past interference (Rogan and Miller, 2006). Such heterogeneous landscapes are defined by land-cover categories that are complicated to be defined spectrally due to low inter-class separability and high intra-class variability.

2. Second, there is a need for algorithms that can be interpreted readily and automated as well as to be easily run with user defined parameters that are simple to adjust.

3. Third, a promising land-cover classification algorithm for large area mapping rely on the capability of the algorithm to work with noisy observations, a complex measurement space, and a few numbers of training data compared to the size of the study area (DeFries and Chan, 2000; Rogan et al., 2008).

A wide range of classification methods have been used to map land cover using remotely sensed data. Classification methods vary from unsupervised algorithms such as K-means klustring to parametric supervised algorithms such as maximum likelihood (Otukei and Blaschke, 2010); to machine learning algorithms such as artificial neural networks (Duro et al, 2012), SVMs (Mountrakis et al., 2011), decision trees (Breiman, 1984; Hua et al., 2017), and ensembles of classifiers (Breiman, 1996).

Dealing with large and complex datasets, machine learning algorithms are more accurate and efficient compared to conventional parametric algorithms, (Rodriguez-Galiano et al., 2012). The usual purpose of land cover classification is to produce a thematic map of the land cover.

Land cover is the material at the ground, such as vegetation, water, soil and man-made structures. (Fisher and Unwin, 2005). The number and kind of land cover classes in the image

* Corresponding author: ali.jamali.65@gmail.com

that can defined vary significantly depending on the sensor resolutions.

An image classification is an image processing technique that according to their spectral signatures, identifies materials in an image. The spectral signature is the reflectance as a function of wavelength where each material has a unique signature, therefore it is called as material classification (NASA, 2013). For environmental research, Landsat images are widely used. Landsat is a set of multispectral satellites developed by the NASA (National Aeronautics and Space Administration of USA), since the early 1970's.

Several pixels and/or regions of pixels of satellite images with known classes as training data are required to predict classes of other regions using the supervised/unsupervised classification model (e.g. Random Forest). Normally, supervised classifications need the user to identify one or more Regions of Interest (ROIs, also Training Areas) for each land cover class identified in the image. ROIs are polygons drawn over homogeneous areas of the image that overlay pixels belonging to the same land cover class.

The spectral signatures (spectral characteristics) of reference land cover classes are defined considering the values of pixels under each ROI having the same Class ID. As a result, the classification algorithm classifies the whole image by comparing the spectral characteristics of each pixel to the spectral characteristics of reference land cover classes.

Assessing the accuracy of land cover classification, in order to identify and measure map errors after the classification process is a usual step. Usually, accuracy assessment is done with the calculation of an error matrix, which is a table that compares map information with reference data for several sample areas (Congalton and Green, 2009).

In this research, for image classification, ML, RF, SVM and a combination of SVM, ML and RF are implemented in R programming language and compared in term of overall accuracy of image classification. Following this introduction, in Section 2, classification methods are discussed. Results of image classification are presented in Section 3. In Section 4, conclusion and future research are discussed.

2. METHODS

Three classification methods including ML, RF and SVM are researched. SVM and RF methods are implemented within R programming language where ML is implemented in QGIS using semi-automatic classification plug-in developed by Luca Congedo.

RF (Breiman, 2001) method is an extension of classification and regression trees (CART; Breiman *et al.*, 1984). RF method is an ensemble learning technique which is increasingly used in land-cover classification using multispectral and hyperspectral satellite sensor imagery. RF creates several trees based on random bootstrapped of the training dataset samples. RF runs random binary trees that creates a subset of the trainings over bootstrapping method, from the initial dataset, a random selection of the training data is selected and implemented to construct the model, out of bag (OOB) is the data which is not included (Catani et al. 2013). The number of trees (ntree), and the number of variables (mtry) are two parameters which are needed to be tuned in a RF method.

SVMs (Vapnik, 1998) uses a simple linear method to the data but in a high-dimensional feature space non-linearly related to the input space, but in practice, it does not use any computations in that high-dimensional space. The combined the state-of-the-art performance and simplicity on many learning problems (regression and classification) has increased the popularity of the SVMs (Leo *et al.*, 2006). SVM is a supervised machine learning technique that is implemented based on the Structural Risk Minimization (SRM) principle and statistical learning theory (Tehrany et al. 2015). SVMs have higher accuracies compared with the traditional approaches but the results rely on the kernel used, choice of parameters for the chosen kernel and the method used to generated SVM (Huang et al., 2002).

A maximum likelihood classification algorithm is a parametric supervised classifier. The algorithm for computing the weighted distance or likelihood D of unknown measurement vector X belong to one of the known classes Mc is based on the Bayesian equation (see equation 1).

$$D = \ln(a_c) - [0.5 \ln(cov_c) - [0.5X - M_c]T(cov_c - 1)(X - M_c)] \quad (1)$$

The probability distributions for the classes are assumed as a form of multivariate normal models (Richards & Jia, 2006). A sufficient number of pixels are needed for each training area for the calculation of the covariance matrix. The discriminant function, described by Richards and Jia (2006), is calculated for every pixel as (see Equation 2):

$$g_k(x) = \ln p(c_k) - \frac{1}{2} \ln |\Sigma k| - \frac{1}{2} (x - y_k)^t \Sigma_k^{-1} (x - y_k) \quad (2)$$

where: c_k = land cover class k;
x = spectral signature vector of an image pixel;
$p(c_k)$ = probability that the correct class is c_k;
$|\Sigma k|$= determinant of the covariance matrix of the data in class Ck;
Σ_k^{-1} = inverse of the covariance matrix;
y_k = spectral signature vector of class k.

The advantage of the ML is that it considers the variance–covariance within the class distributions and for normally distributed data, the ML performs better than the other known parametric classifies (Erdas, 1999). For data with a non-normal distribution, ML may have unsatisfactory results.

3. RESULTS

Shiraz is located in the south of Iran which is built in a green plain at the foot of the Zagros Mountains, 1,500 meters (4,900 feet) above sea level. Shiraz has a considerable number of gardens which due to clime change (i.e. droughts) and population growth in the city, many of these gardens may be lost for new urban city developments. Although the Municipality has taken some measures to preserve these gardens, land cover monitoring is a key factor to monitor preservation of garden regions over time.

As seen in Figure 1, to visualize the Shiraz urban (build-up) growth, Normalized Difference Build-up Index (NDBI) is calculated (see Equation 3).

$$NDBI = (B - SWIR)/(B + SWIR) \quad (3)$$

Where SWIR is Short Wave Infra-Red band and;
B is Blue band.

Most of classification methods take a formula such as $Y = X_1 + X_2$, to find dependent and independent variables where Y is a function of X_1 and X_2. In case of satellites images classification, several levels of classes (i.e. build-in, wet lands, crop) are estimated by a combination of several spectral signature of different bands with/without Normalized Vegetation Difference Index (NDVI) (see Equation 4).

$$NDVI = (NIR - R)/(NIR + R) \quad (4)$$

Where NIR is Near Infra-Red band and;
R is Red band.

Considering six bands plus NDVI for Landsat 8, in R programming language, the classification formula is written by (see Equation 5):

classes ~ band2 + band3 + band4 + band5 + band6 + band7 + NDVI (5)

Landsat images are required to be pre-processed for sensor, solar, atmospheric and topographic effects (Yang et al., 2017). In this research, semi-automatic classification plug-in is used for Dark Object Subtraction (DOS) which is an image-based atmospheric correction. The data for image classification is from the Landsat 8 OLI satellite belonging to 22nd August 2018 data set of Shiraz city in WGS 84 / UTM zone 39N (Figure 2). Images classification using RF (see Figure 3), SVM (see Figure 4), ML (see Figure 5) and a combination of three methods (see Figure 6) are presented.

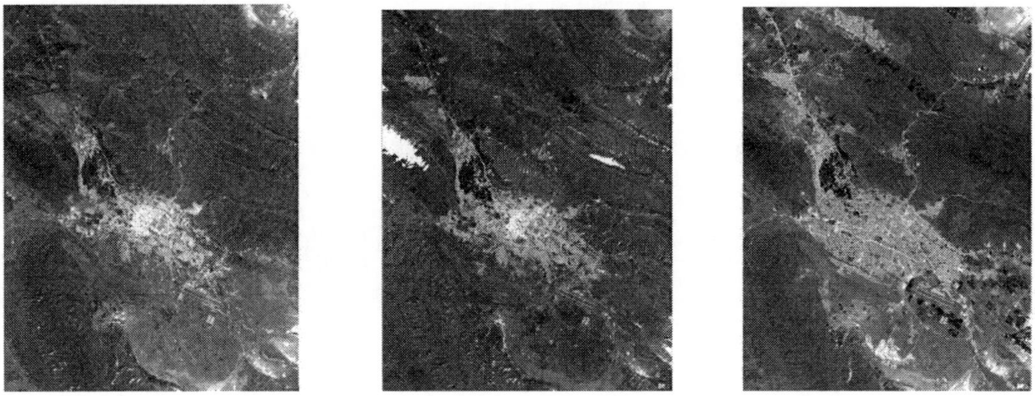

Figure 1. Normalized Difference Build-up Index (NDBI) for Shiraz city in 1990 (left picture, Landsat 5 image), 2000 (middle picture, Landsat 7 image) and 2018 (right picture, Landsat 8 OLI image).

https://doi.org/10.5194/isprs-archives-XLII-3-W7-25-2019

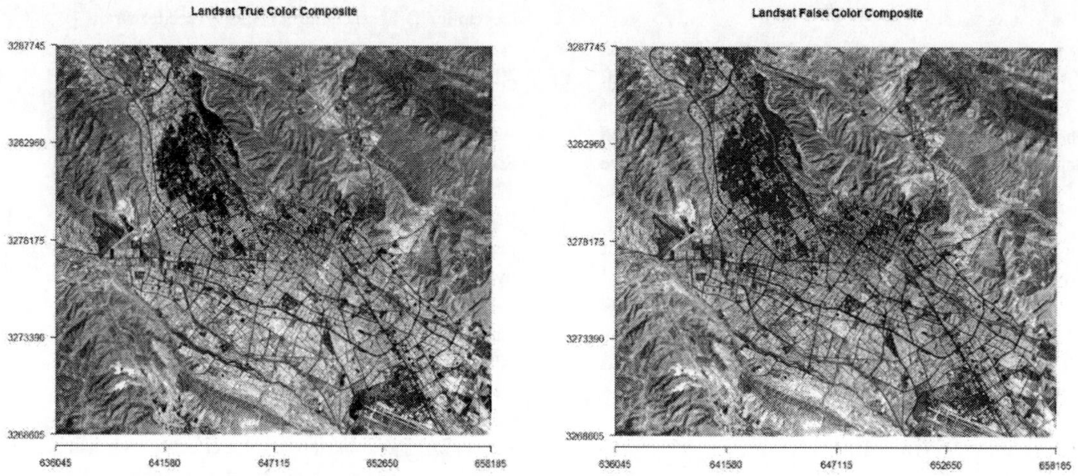

Figure 2. Shiraz city.

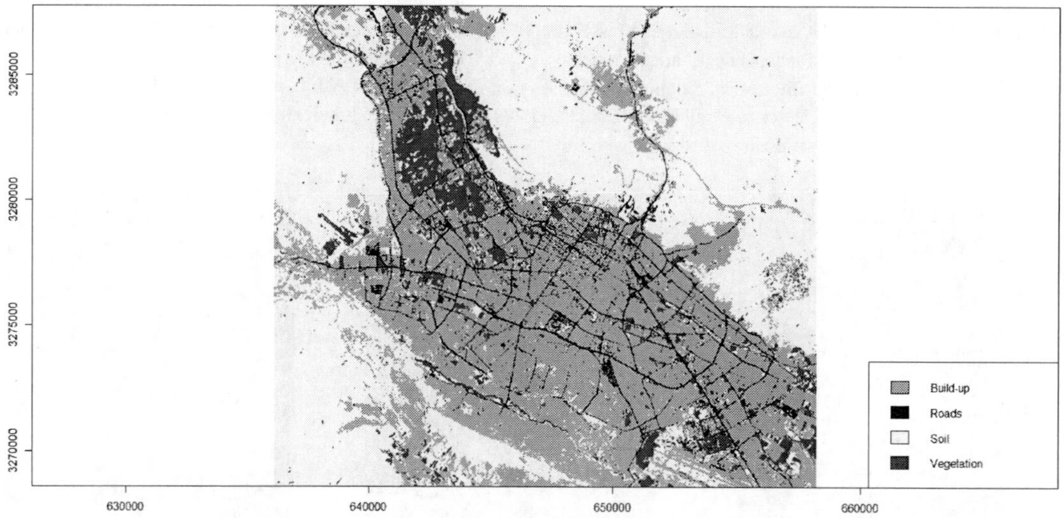

Figure 3. Image classification using Random Forest classification technique within R programming language.

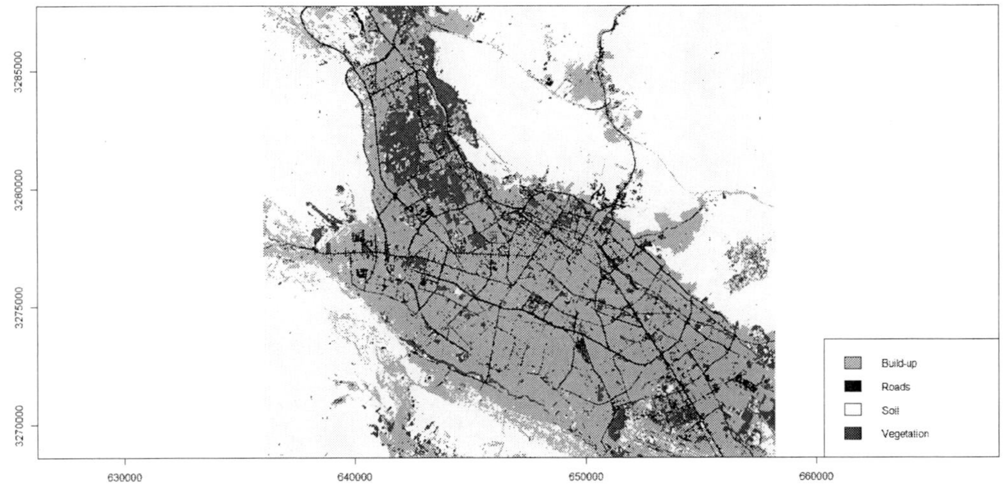

Figure 4. Image classification using SVM classification technique within R programming language.

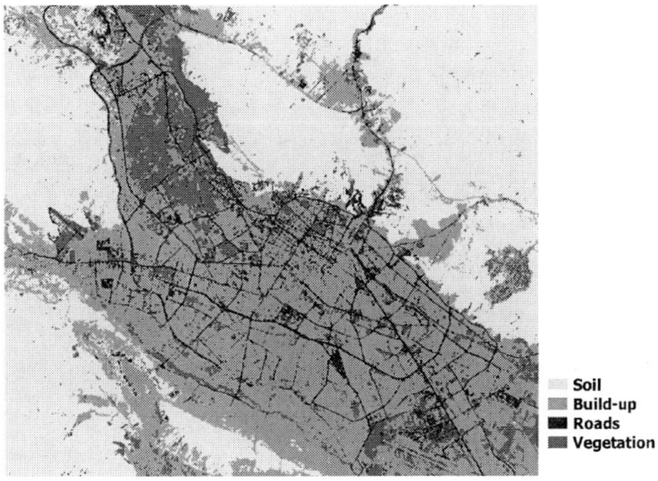

Figure 5. Image classification using ML classification technique with semi-automatic classification plug-in developed by Luca Congedo.

https://doi.org/10.5194/isprs-archives-XLII-3-W7-25-2019

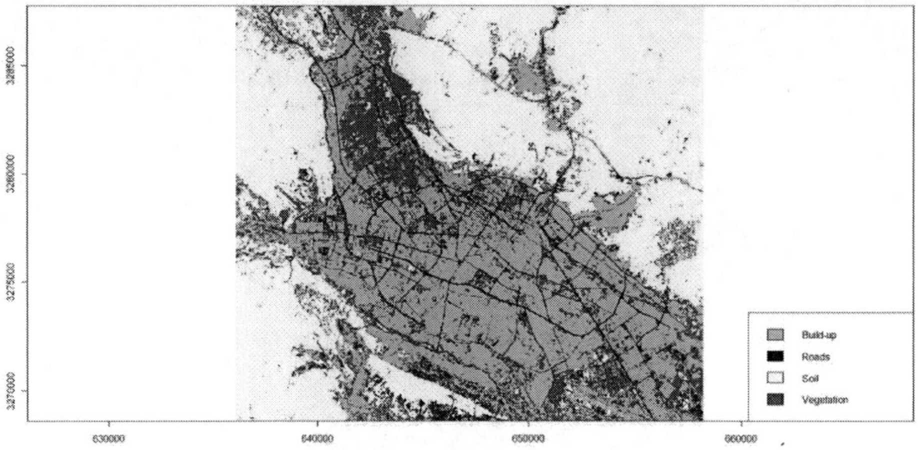

Figure 6. Image classification using a combination of SVM, ML and RF classification techniques within R programming language.

Overall accuracy of image classifications is presented in Tables 1 to 4. Images are classified by four materials including build-up, roads, soil and vegetation regions to evaluate performance of three image classification methods. RF, SVM and ML classification methods have overall accuracies of 99.83, 99.93 and 99.7 percent respectively for the evaluation dataset.

RF	Build-up	Roads	Soil	Vegetation
Build-up	2591	2	17	0
Roads	2	359	0	0
Soil	13	1	19485	0
Vegetation	0	7	0	1895
Overall accuracy	**99.83**			

Table 1. Accuracy of predicted materials by RF classification technique.

SVM	Build-up	Roads	Soil	Vegetation
Build-up	2603	2	4	0
Roads	2	360	0	0
Soil	1	0	19498	0
Vegetation	0	7	0	1895
Overall accuracy	**99.93**			

Table 2. Accuracy of predicted materials by SVM classification technique.

ML	Build-up	Roads	Soil	Vegetation
Build-up	2600	1	44	0
Roads	5	364	0	0
Soil	1	0	19440	0
Vegetation	1	4	18	1895
Overall accuracy	**99.70**			

Table 3. Accuracy of predicted materials by ML classification technique.

An advantage of R programming language is possibility of combining different classification methods considering a fit-for-purpose algorithm within an efficient, feasible and easy-to-use platform. Overall accuracy of the combined method is 99.71 percent for the evaluation dataset (see Table 4).

Combined	Build-up	Roads	Soil	Vegetation
Build-up	2600	3	8	0
Roads	6	360	48	0
Soil	0	2	19446	0
Vegetation	0	4	0	1895
Overall accuracy	**99.71**			

Table 4. Accuracy of predicted materials by a combination of SVM, ML and RF classification techniques.

As the main goal is the identification of garden regions in Shiraz city, the combined method for extraction of vegetation information shows better results compared to RF, ML and SVM methods for the evaluation dataset.

4. CONCLUSIONS

Free and commercial Earth Observation (EO) satellites sensor data are a key factor for large area environmental monitoring. Due to several climate change phenomena (e.g. increase of temperature due to green house gasses) in recent years and their impact on the land cover change and vice versa; the effect of land cover changes on earth climate, image classification for large area environments is a necessity.

R programming language is a powerful and efficient platform for EO researchers to implement their algorithm to monitor and model recent earth climate changes over land cover. A fit-for-purpose algorithms can be designed for certain applications such as vegetation extraction, Flood monitoring

https://doi.org/10.5194/isprs-archives-XLII-3-W7-25-2019 | © Authors 2019. CC BY 4.0 License.

and man-made (build-up) area estimation. Several classification algorithms can be combined to extract certain information to satisfy requirements of an end-user. In the case of the Shiraz city, to monitor preservation of garden regions, an algorithm just for identification of trees over other materials (e.g. water and man-made area) with a high accuracy is required to be constructed.

REFERENCES

Bala, G., Caldeira, K., Wickett, M., Phillips, T., Lobell, D., Delire, C., Mirin, A., 2007. Combined climate and carbon-cycle effects of large-scale deforestation. Proceedings of the National Academy of Sciences 104 (16), 6550–6555.

Betts, R., Falloon, P., Goldewijk, K., Ramankutty, N., 2007. Biogeophysical effects of land use on climate: model simulations of radiative forcing and large-scale temperature change. Agricultural and Forest Meteorology 142 (2–4), 216–233.

Bonan, G., 2008. Forests and climate change: forcings, feedbacks, and the climate benefits of forests. Science 320 (5882), 1444–1449.

Breiman, L., 1984. Classification and Regression Trees. Chapman & Hall/CRC.

Breiman, L., 2001. Random forests. *Machine learning*, *45*(1), 5-32.

Breiman, L., 1996. Bagging predictors. Machine Learning 24 (2), 123–140.

Catani F, Lagomarsino D, Segoni S, Tofani V., 2013 Landslide susceptibility estimation by random forests technique: sensitivity and scaling issues. Nat Hazards Earth Syst Sci 13:2815–2831.

Congalton, R. and Green, K., 2009. Assessing the Accuracy of Remotely Sensed Data: Principles and Practices. Boca Raton, FL: CRC Press.

DeFries, R.S., Chan, J.C.-W., 2000. Multiple criteria for evaluating machine learning algorithms for land cover classification from satellite data. Remote Sensing of Environment 74 (3), 503–515.

Duro, D. C., Franklin, S. E., & Dubé, M. G., 2012. A comparison of pixel-based and object-based image analysis with selected machine learning algorithms for the classification of agricultural landscapes using SPOT-5 HRG imagery. *Remote Sensing of Environment*, *118*, 259-272.

Erdas Inc., 1999. Erdas Field Guide. Erdas Inc., Atlanta, Georgia.

Fisher, P. F. and Unwin, D. J., eds. 2005. Representing GIS. Chichester, England: John Wiley & Sons.

Foody, G.M., 2002. Status of land cover classification accuracy assessment. Remote Sensing of the Environment 80, 185–201.

Gevrey, M., Dimopoulos, I., & Lek, S., 2003. Review and comparison of methods to study the contribution of variables in artificial neural network models. *Ecological modelling*, *160*(3), 249-264.

Grippa, T., Georganos, S., Zarougui, S., Bognounou, P., Diboulo, E., Forget, Y., ... & Wolff, E., 2018. Mapping Urban Land Use at Street Block Level Using OpenStreetMap, Remote Sensing Data, and Spatial Metrics. *ISPRS International Journal of Geo-Information*, *7*(7), 246.

JARS, 1993. Remote Sensing Note. Japan Association on Remote Sensing. Available at http://www.jars1974.net/pdf/rsnote_e.html.

Han, S., Ren, F., Wu, C., Chen, Y., Du, Q., & Ye, X., 2018. Using the TensorFlow Deep Neural Network to Classify Mainland China Visitor Behaviours in Hong Kong from Check-in Data. *ISPRS International Journal of Geo-Information*, *7*(4), 158.

Hua, L., Zhang, X., Chen, X., Yin, K., & Tang, L., 2017. A Feature-Based Approach of Decision Tree Classification to Map Time Series Urban Land Use and Land Cover with Landsat 5 TM and Landsat 8 OLI in a Coastal City, China. *ISPRS International Journal of Geo-Information*, *6*(11), 331.

Huang, C., Davis, L.S., Townshed, J.R.G., 2002. An assessment of support Vector Machines for Land cover classification. International Journal of Remote sensing 23, 725–749.

Leo, B., Friedman, J. H., Olshen, R. A., & Stone, C. J., 1984. Classification and regression trees. *Wadsworth International Group*.

Mountrakis, G., Im, J., Ogole, C., 2011. Support vector machines in remote sensing: a review. ISPRS Journal of Photogrammetry and Remote Sensing 66 (3), 247–259.

Otukei, J. R., & Blaschke, T., 2010. Land cover change assessment using decision trees, support vector machines and maximum likelihood classification algorithms. *International Journal of Applied Earth Observation and Geoinformation*, *12*, S27-S31.

NASA, 2013. Landsat 7 Science Data User's Handbook. Available at http://landsathandbook.gsfc.nasa.gov.

Rodriguez-Galiano, V. F., Ghimire, B., Rogan, J., Chica-Olmo, M., & Rigol-Sanchez, J. P., 2012. An assessment of the effectiveness of a random forest classifier for land-cover classification. *ISPRS Journal of Photogrammetry and Remote Sensing*, *67*, 93-104.

Rogan, J., Miller, J., 2006. Integrating GIS and remotely sensed data for mapping forest disturbance and change. In: Franklin, M.W.A.S. (Ed.), Understanding Forest Disturbance and Spatial Pattern: Remote Sensing and GIS Approaches. CRC Press, Boca Raton, FL, pp. 133–172.

Rogan, J., Franklin, J., Stow, D., Miller, J., Woodcock, C., Roberts, D., 2008. Mapping land-cover modifications over large areas: a comparison of machine learning algorithms. Remote Sensing of Environment 112 (5), 2272–2283.

Skole, D.L., 1994. Data on global land cover change: acquisition assessment and analysis. In: Turner, II, W.B. (Ed.), Changes in Land Use and Land Cover: A Global Perspective. Cambridge University Press, Cambridge, pp. 437–471.

Tehrany MS, Pradhan B, Mansor S, Ahmad N., 2015. Flood susceptibility assessment using GIS-based support vector machine model with different kernel types. Catena 125:91–101. doi:10.1016/j.catena.2014.10.017.

Vitousek, P.M., 1994. Beyond global warming: ecology and global change. Ecology 75, 1861–1876.

Young, N. E., Anderson, R. S., Chignell, S. M., Vorster, A. G., Lawrence, R., & Evangelista, P. H. (2017). A survival guide to Landsat preprocessing. *Ecology*, *98*(4), 920-932.

AGROFORESTRY TREE DENSITY ESTIMATION BASED ON HEMISPHERICAL PHOTOS & LANDSAT 8 OLI/TIRS IMAGE: A CASE STUDY AT CIDANAU WATERSHED, BANTEN-INDONESIA

R. N. Khairiah[1], L. B. Prasetyo[2]*, Y. Setiawan[2]

[1]Graduate Program of Tropical Biodiversity Conservation, Department of Forest Resources Conservation Department, Forestry Faculty, Bogor Agricultural University- rahmi_khairiah10@apps.ipb.ac.id, [2] Department of Forest Resources Conservation Department, Forestry Faculty, Bogor Agricultural University (lbprastdp@apps.ipb.ac.id)

Commission III, WG III/10

Keywords: FCD, LAI, Hemispherical view, Tree density, topographical correction

ABSTRACT:

The Cidanau watershed is the only watershed in Indonesia that implements Payment for Environmental Services (PES) for farmers who can maintain tree/stand density of 500 trees/hectare on their land. Payments are made upon the verification on the field by the project supervisor. This method requires a lot of time and costly, so it is necessary to build more efficient indirect methods, including using satellite imagery or camera data. The aim of this study is to understand Landsat OLI 8 and hemispherical photo can estimate tree density in the farmer's agroforestry stand. To obtain tree density, the number of trees with diameter more than 10 cm in 50 plots (50 m x 50 m) were counted. Some predictor variables were utilized, such as Leaf Area Index (LAI) based on hemispherical photos, Normalized Difference Vegetation Index (NDVI), Forest Cover Density (FCD), as well as NDVI and FCD which were enhanced with topographic correction. The imagery used was Landsat 8 OLI acquired on July 5, 2015, with Path/Row 123/64. The relationship between tree density and predictor variables was done using linear regression analysis. Prior to regression analysis, normality (Kolmogorov Smirnov/K-S), heteroscedasticity (Glejser test) and auto correlation (Durbin Watson) test were performed. The results of the analysis showed that tree density was estimated better with hemispherical photos-based LAI, with determination coefficient of 80.6%. Meanwhile, estimation using NDVI and FCD has lower determination coefficient. Even though, the use of topographic correction had been able to increase the determination coefficient of the regression relationship between tree density and FCD, from 4.64% to 35.18%.

1. INRODUCTION

Cidanau River is one of the rivers in Banten Province which is used for domestic and industrial drinking water, which is managed by a private water company PT. Krakatau Tirta Industri (PT. KTI). The communities in this watershed are mostly rice farmers in flat areas and agroforestry farmers in hilly areas. Agroforestry is a farming technique that combines trees with food crops (Nair, 1993). Continuous water supply from the river depends on the ability of the watershed to store water throughout the year. This depends greatly on the conditions of land cover. The better the vegetation closure, the better the watershed will store water. This condition is indicated by the coefficient value of river regime, which is a ratio between maximum and minimum discharge (Qmax/Qmin), that is not greater than 15 (Pantouw et al., 2013).

To maintain extensive tree vegetation closure, PT. KTI runs a Payment for Environmental Service (PES) program, where farmers will get a payment if they can maintain a tree density of 500 trees/hectare in their agroforestry land. Payment will be made after verification in the field by the assessor/project supervisor. Field work assessment usually takes a long time and high costs. Therefore, efficient indirect methods need to be found, including using satellite image data and cameras. This research aimed at understanding how Forest Cover Density (FCD) and hemispherical photos-based Leaf Area Index (LAI) can estimate tree density in the farmer's agroforestry stand.

Forest Canopy Density (FCD) is a quantitative value that represents the degrees of forest density and expressed by the percentage. It is an important variable related to the forest degradation, as a basic information of possible management intervention (Azizi et al., 2008). Estimation of FCD using satellite imagery has been developed by Rikimaru (Rikimaru et al., 2002) based on 4 types of indices, namely temperature, shadow, vegetation and open land index based on satellite image data. FCD has been tested in many aspects of forest management, including estimating forest density (Banerjee, et al., 2014; Anand et al., 2018), and basal area and predominant height (Baynes, 2004). Meanwhile, Muhammad et al. (2014) and Tohir, et al. (2014) studied pine forests and community forests with teak dominance, respectively, and both results showed a good accuracies. In both cases the tree stand structure was relatively uniform, and until now there is no literature yet on the use of FCD in mixed/heterogeneous ecosystems, such as agroforestry ecosystem. Moreover, most of the Landsat images for FCD identification were not topographically corrected. The current research is the first attempt to calculate tree density using FCD for the agroforestry ecosystem with topographically corrected Landsat images.

Hemiview photos have been used to estimate Leaf Area Index (LAI) (Anne et al., 2006; Lhotka & Loewenstein, 2006; Chen et al., 1991), but the use on estimating tree density in agroforestry stand is very limited. In Indonesia, estimations of stand density using Hemiview photos carried out by Muhammad et al. (2014) in pine stands and Tohir, et al. (2014) in community plantation forests dominated by teak trees. The aim of the research is understanding how

Normalized Vegetation Index (NDVI), Forest Cover Density (FCD) and hemispherical photos-based Leaf Area Index (LAI) can estimate tree density in the farmer's agroforestry stand. Meanwhile the objectives of the research are as follows : (1) to calculate and compare stand density using NDVI, FCD and Hemispherical photos-based LAI for the agroforestry ecosystem with topographically corrected Landsat images and (2) to clarify the mathematical relationship between tree density values with NDVI, FCD and Hemispherical photos-based LAI

2. MATERIAL AND METHOD

2.1 Study site

The study was conducted in the Cidanau Watershed, Serang District and Pandeglang District, Banten Province, Indonesia. Geographically, the Cidanau watershed is located between 07o 07 '30" - 06 o 18 '00" South Latitude and 105o 49 '00" -106 o 04 '00" East Longitude. The Cidanau watershed has an area of 22,620 hectares. The watershed landscape is mountainous at upstream, flat in the middle and relatively gentle slope a the downstream (Figure 1). Land cover of upstream is forests and various types of agroforestry ecosystem. In the middle part is food crops and in the downstream is agroforestry. The majority of the people living in the area are farmers.

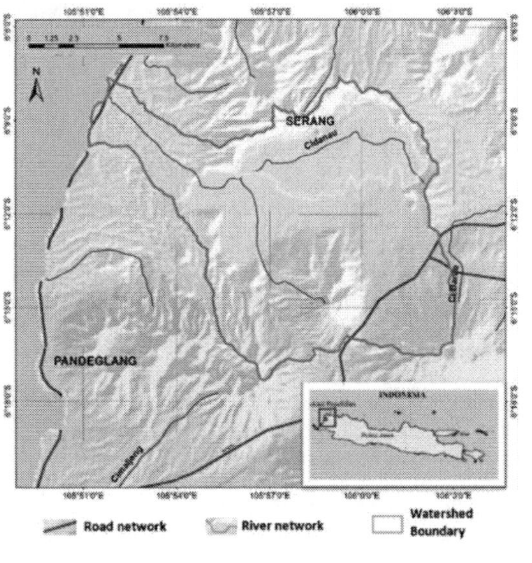

Figure 1. Study area: Cidanau Watershed

2.2 Methods

2.2.1 Survey plot and measurement methods

Survey plot locations were distributed purposively based on 10 categories of FCD, ranging from low FCD (0-10%) to very high FCD (90-100%). A total of 50 plots (50 m x 50 m) were established for hemispherical photographs and tree stand density measurement. In addition, 50 additional plots from secondary data of the previous research were used for tree density survey. Hemispherical view photographs were taken from the centre of the plot at position of ± 1.5 m height from the ground, using DSLR camera mounted with fisheye lens. Tree species and number of trees with diameter more than 10 cm within the plot were counted.

2.2.2 Topographic correction

The Landsat 8 OLI/TIRS Path/Row 123/64, taken on 5 July 2015 image was topographically corrected based on Tan algorithm (Tan *et al.*, 2013), which also adopted by Hudjimartsu (2017) to reduce the effect of sun illumination difference. Illumination condition and corrected reflectance model were calculated based on equation as follows:

$$Ic = \cos\phi \cos slope + \sin\phi \sin slope \cos(\theta\text{-}aspect) \quad (1)$$

$$\lambda_{corrected} = \lambda_{uncorrected} - \beta(ic-\phi) \quad (2)$$

Where :
ic : Illumination condition, ϕ : Zenith angle, θ : Azimuth angle, β is the slope of the linear regression between illumination condition of each pixel and reflectance of the same object.

2.2.3 Forest Cover Density (FCD)

Forest Cover Density algorithm was develop by Rikimaru (Rikimaru *et al.*, 2002). The FCD is representing the proportion of tree cover and land surface under the canopy. Its value range from 0 to 100%. Value of 0% means there is no tree canopy, conversely 100% means all land surface is covered by tree canopy. In this research, the FCD was estimated topographical corrected Landsat images, by using FCD mapper. Two FCD calculations were performed, namely before and after topographical correction. Topographic correction will eliminate the slope and aspect effect, because the pixel reflectance has been corrected with illumination conditions. The illumination correction was more accurate, since it was considering sun angle and sun azimuth for every pixel of the image, rather than sun angle and sun azimuth of midpoint of the scene written in the Landsat image metadata.

2.2.4 Leaf Area Index (LAI) and hemispherical view

Leaf Area Index is projected area of green canopy per unit of horizontal ground surface area (Casella et al., 2013). Hemispherical view photographs were taken by Nikon camera mounted with fisheye lens. The LAI was calculated by using Hemiview canopy analysis software.

Figure 2. Hemispherical view acquisition and its photograph

https://doi.org/10.5194/isprs-archives-XLII-3-W7-33-2019 | © Authors 2019. CC BY 4.0 License.

2.2.5 Normalized difference vegetation index (NDVI)

NDVI was calculated based on an equation as follows:

$$NDVI = \frac{NIR - RED}{NIR + RED} \qquad (3)$$

Where :
NDVI = Normalized Difference Vegetation Index
NIR = Near Infra red band (Landsat 8 OLI, 0.851 - 0.879 μm)
Red = Red band (Landsat 8 OLI, 0.636 - 0.673 μm)

2.2.6 Statistical Analysis

Linear regression analysis between tree stand density with FCD and LAI were conducted. Prior to this analysis, classical assumption was checked, including the assumptions of normality (Kolmogorov Smirnov/K-S), heteroscedasticity (Glejser test) and auto correlation (Durbin Watson) by using software IBM SPSS Statistics 22.

3. RESULTS

3.1 Topographic correction

Topographic correction could reduce the influence of illumination differences as presented in Figure 3. Further observation to the pixel reflectance of every bands in the Landsat 8 OLI showed that the topographic correction reduced the effect of different illumination. Before topographic correction, the pixel reflectance of the same object increased by the increasing illumination condition. Meanwhile, after the correction, the same object most likely have the same spectral reflectance. Figure 4 to Figure 9 presents the response of pixel reflectance some bands of Landsat 8 OLI due to different illumination before and after topographic correction.

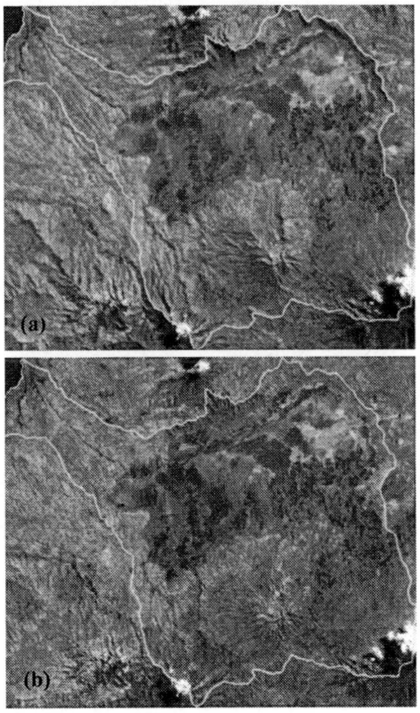

Figure 3. Landsat 8 OLI image comparison (a) before and (b) after topographic correction

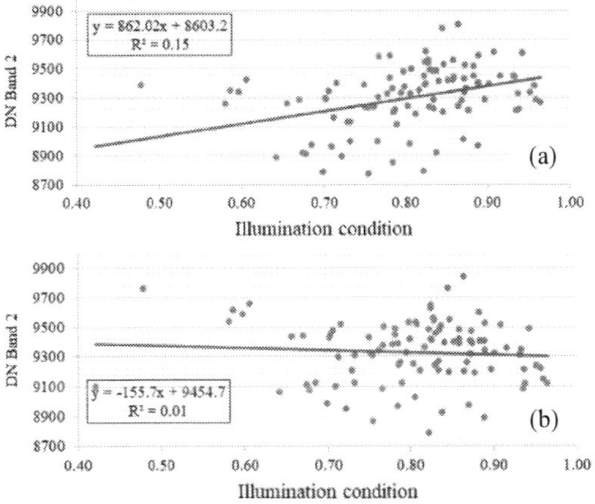

Figure 4. Reflectance of band 2, (a) before and (b) after topographic correction

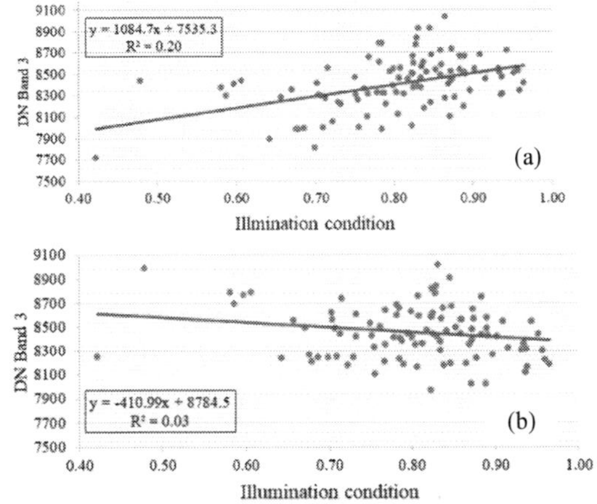

Figure 5. Reflectance of band 3, (a) before and (b) after topographic correction

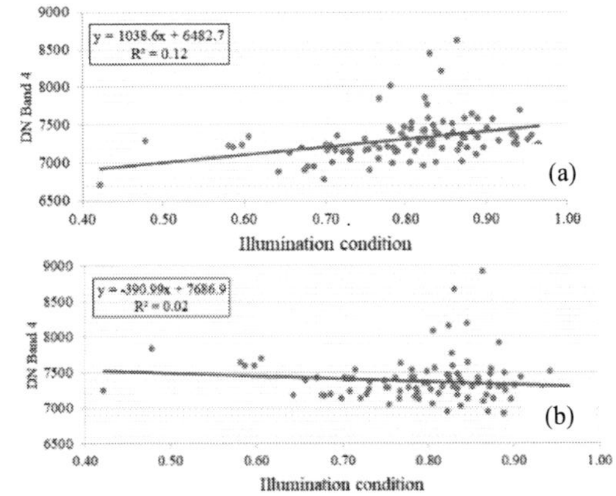

Figure 6. Reflectance of band 4, (a) before and (b) after topographic correction

This contribution has been peer-reviewed.
https://doi.org/10.5194/isprs-archives-XLII-3-W7-33-2019 | © Authors 2019. CC BY 4.0 License.

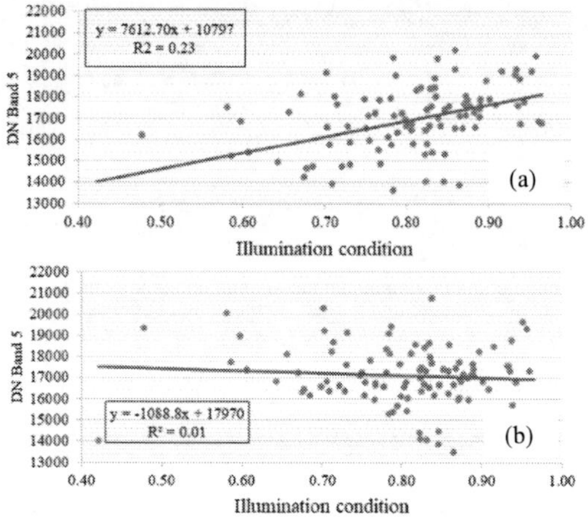

Figure 7. Reflectance of band 5, (a) before and (b) after topographic correction

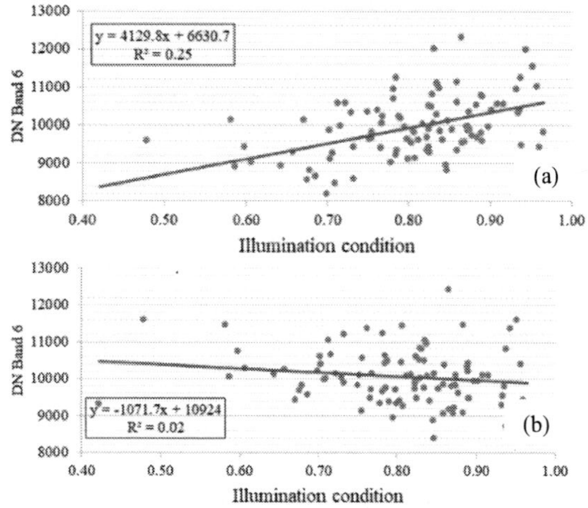

Figure 8. Reflectance of band 6, (a) before and (b) after topographic correction

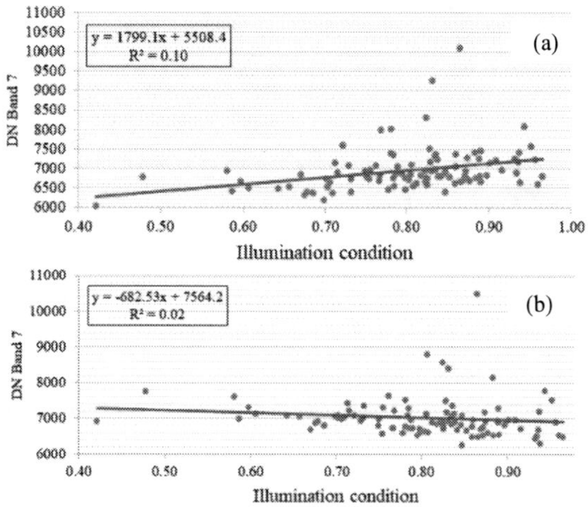

Figure 9. Reflectance of band 7, (a) before and (b) after topographic correction

3.2 NDVI before and after topographic correction

Visually, topographic correction was not influence the NDVI significantly (Figure 10). It could be explained also by the value of statistic parameter of the images, whereas the value was only slightly changing (Table 2). Since NDVI was based on ratio bands, the topographic correction did not have much effect because it's avoiding each other.

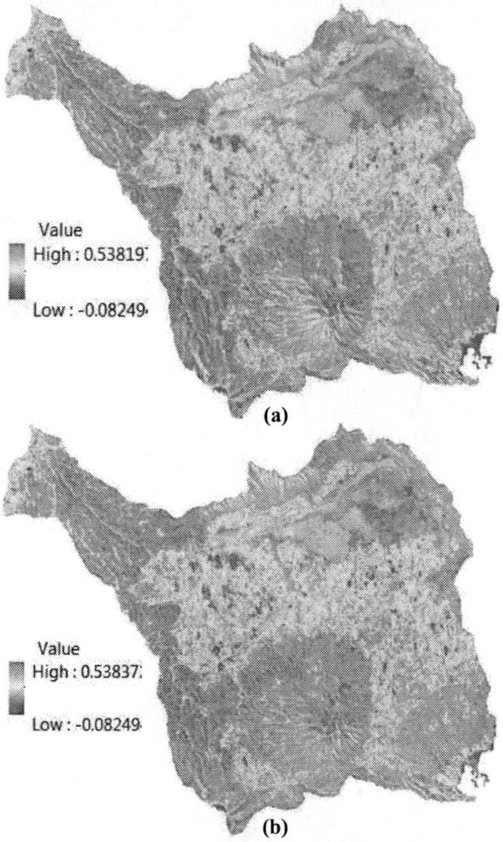

Figure 10. Value of NDVI, (a) before and (b) after topographic correction

Table 1.Statistical parameter comparison of NDVI before and after topographic correction

NDVI	Before topographic correction	After topographic correction
Minimum	-0.0825	-0.0825
Maximum	0.5382	0.5384
Average	0.201	0.203
Std. Dev.	0.184	0.186

3.3 FCD before and after topographic correction

Since the topographic correction reduce the influence of slope & aspect, the correction also affecting the value of FCD (Figure 11). The FCD before topographic correction tended to be higher in comparison with FCD after correction. This is shown in the Table 2 that the maximum FCD increased, while the average, median and standard deviation of FCD decreased.

Table 2.Statistical parameter comparison of FCD before and after topographic correction

FCD	Before Correction	After Correction
Minimum	0%	0%
Maximum	91%	94%
Average	55.17%	50.09%
Std. Dev.	27.92%	23.95%

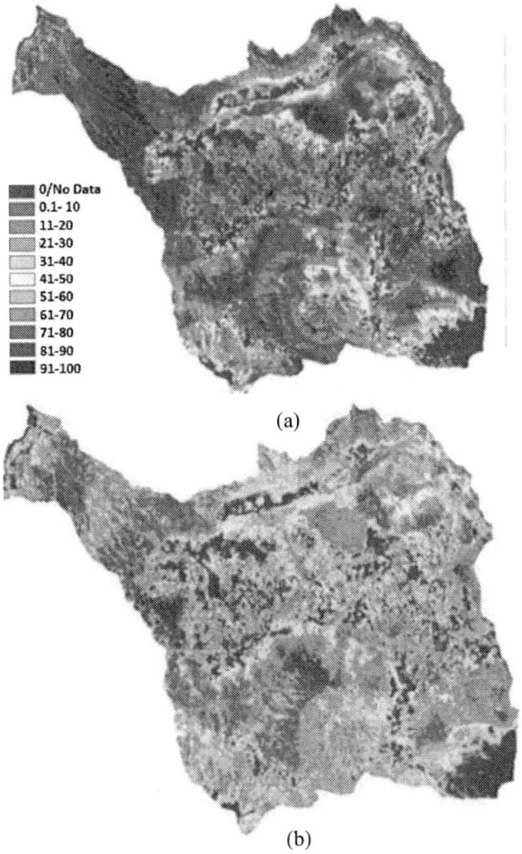

Figure 11. FCD value, (a) before and (b) after topographic correction

3.4 Relation of Tree density, FCD and LAI

Upon classic assumption check of linear regression with prior normality, heteroscedasticity, and autocorrelation, three linear regression equations were performed.

3.4.1 Linear regression between tree density and NDVI before and after topographical correction

Even though the correction of topographic have not much influence on the NDVI, the correction could improve the coefficient determination (R^2) of the regression analysis from 0.1576 to 0.2962, between tree density and NDVI (Figure 12).

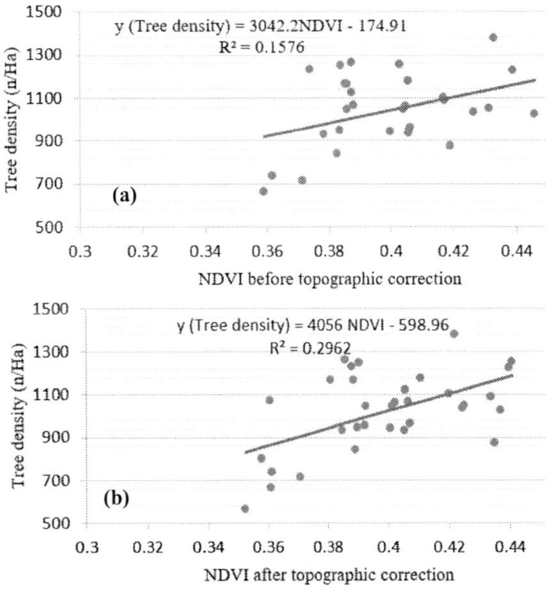

Figure 12. Regression between tree density and NDVI, (a) before and (b) after topographic correction

3.4.2 Linear regression between tree density and FCD before and after topographic correction.

Figure 13 shows the regression between tree density and FCD before topographic correction. The graphic shows a negative relationship between FCD and tree density, moreover, the relation was very weak (R^2 = 0.046.). This is contrary to the results of the research conducted by Tohir, et al. (2014) and Muhammad, et al. (2014), whereas the FCD in line with the tree density, regardless the difference stand structure.

After topographic correction, the relationship between tree density and FCD changed to positive, the more FCD increases, the higher the density of the tree. In addition, the coefficient of determination (R^2) also increases from 0.0464 to 0.3518 (Figure 13).

3.4.3 Linear regression between tree density and LAI

In comparison of employing FCD, the estimation of tree density using LAI was better. Figure 14 shows a very close relationship between tree density and LAI with a determination coefficient (R^2) reaching more than 80%.

4. CONCLUSION

Topographic correction can reduce the effects of different illumination and could improve the quality of satellite image data. Therefore, it can increase the relationship between tree density and FCD as well as NDVI.

This study also concluded that Hemispherical photos-based LAI is a better variable than NDVI and FCD for estimating tree density in agroforestry ecosystem. Therefore, this method is recommended as a proxy to calculate tree density in complex stand structure, such as agroforestry ecosystem.

https://doi.org/10.5194/isprs-archives-XLII-3-W7-33-2019

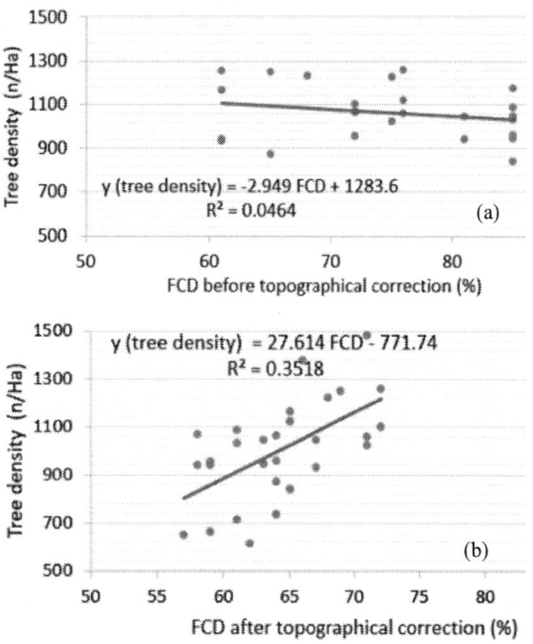

Figure 13. Regression between tree density and FCD, (a) before and (b) after topographic correction

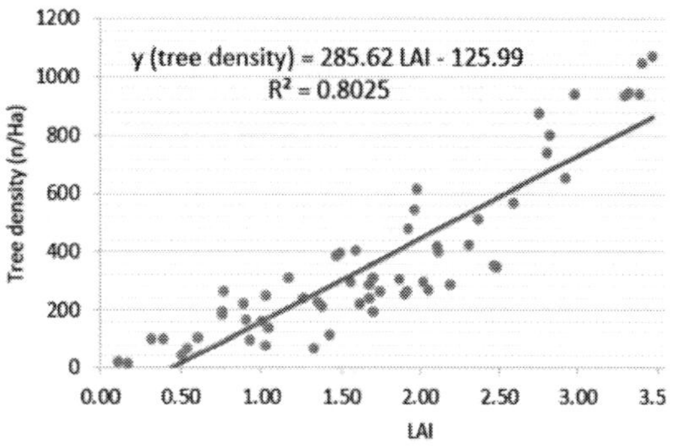

Figure 14. Regression between tree density and LAI derived from hemispherical view

ACKNOWLEDGEMENT

The research are supported by the UK Space Agency (UKSA) International Partnership Program (IPP) under the Forests2020 program coordinated by Ecometrica, Ltd. The Forests 2020 is a collaborative program to advance Earth Observation applications to forests monitoring.

REFERENCES

Anand, A., Singh, S. K., & Kanga, S. 2018. Estimating the change in Forest Cover Density and Predicting NDVI for West Singhbhum using Linear Regression. *International Journal for Environmental Rehabilitation and Conservation*, IX(1), 193–203. https://doi.org/ 10.31786/ 09756272.18.9. 1.125

Anne C.S. Fiala, Garman S.L., Gray A.N. 2006. Comparison of five canopy cover estimation techniques in the western Oregon Cascades. *For Ecol Manage.* 2006;232:188-197. doi:10.1016/j.foreco.2006.05.069

Azizi Z, Najafi A, Sohrabi H. 2008. Forest canopy density estimating, using satelite images. In: *The International Achives of the Photogrammetry, Remote Sensing and Spatial Information System*. ; 2008:1127-1130.

Banerjee, K., Panda, S., Bandyopadhyay, J., & Jain, M. K. 2014. Forest Canopy Density Mapping Using Advance Geospatial Technique. *International Journal of Innovative Science, Engineering & Technology*, 1(7), 358–363.

Baynes, J. 2004. Assessing forest canopy density in a highly variable landscape using Landsat data and FCD Mapper software. *Australian Forestry*, 67(4), 247–253.

Casella E, Disney M, Morison J, Mckay H. 2013. tLiDAR methodologies can overcome limitations in estimating forest canopy LAI from conventional hemispherical photograph analyses. In: *Proceedings of the 7th International Conference on Functional-Structural Plant Models, Saariselkä*, Finland, 9 - 14 June 2013. ; 2013:52-54.

Chen J.M., Black T, Adams RS. 1991. Evaluation of hemispherical photography for determining plant area index and geometry of a forest stand. *Agric For Meteorol.* 1991;56:129-143.

Hudjimartsu S, Prasetyo L.B., Setiawan Y, Suyamto D. 2017. Illumination Modelling for Topographic Correction of Landsat 8 and Sentinel-2A Imageries. In: *Proceedings - UKSim-AMSS 11th European Modelling Symposium on Computer Modelling and Simulation, EMS 2017.* ; 2017. doi:10.1109/EMS.2017.27

Lhotka J.M, Loewenstein E.F. 2006. Indirect measures for characterizing light along a gradient of mixed-hardwood riparian forest canopy structures. *For. Ecol Manage.* 2006;226:310-318. doi:10.1016/ j.foreco.2006. 01.043

Muhammad, A., Prasetyo, L. B., & Kartono, A. P. 2014. Pemetaan Perubahan Forest Canopy Density di KPH Kuningan. In *Seminar Nasional Penginderaan Jauh 2014.Bogor-Indonesia*

Nair P. 1993. *An Introduction to Agroforestry*. London: ICRAF-Kluwer Academic Publisher; 1993

Pantouw, J. P., Limantara, L. M., Bisri, M., & Rispiningtati. 2013. Ratio Between Maximum and Minimum Discharge (Qmax/Qmin) as the Anticipated Indicator of River Disaster in 30 Watersheds of Indonesian. *World Applied Sciences Journal*, 25(7), 1044–1052. https://doi.org/10.5829/idosi.wasj.2013. 25.07. 13379

Rikimaru, A., P.S.Roy, & S.Miyatake. 2002. Tropical forest cover density mapping. *Tropical Ecology*, 43(1), 39–47.

Tan, B., Masek, J. G., Wolfe, R., Gao, F., Huang, C., Vermote, E.F.,Ederer, G. 2013. Remote Sensing of Environment Improved forest change detection with terrain illumination corrected Landsat images. *Remote Sensing of Environment,* 136, 469–483. https://doi.org/10.1016/j.rse.2013.05.013

Tohir N.R, Prasetyo L.B, Kartono A.P. 2014. Pemetaan perubahan kerapatan kanopi hutan di hutan rakyat, Kabupaten Kuningan, Jawa Barat. In: *Seminar Nasional Penginderaan Jauh* 2014. ; 2014:322-341.

ANALYSIS OF RELATIONSHIP BETWEEN VEGETATION DISTRIBUTION AND LAND PRICE USING MULTITEMPORAL DATA

K. Kumagai [1], H. Uematsu[2]

[1] Dept. of Civil and Environmental Engineering, Setsunan University, 17-8 Ikeda-Nakamachi, Neyagawa, Osaka 572-8508, Japan - kumagai@civ.setsunan.ac.jp
[2] Kyoto City Board of Education, Kyoto, Japan - hitoshi.uematsu@gmail.com

Commission III, WG III/10

KEY WORDS: NDVI, Spatial Analysis, GWR, Land Price, Multitemporal Data

ABSTRACT:

Japanese cities are facing a rapidly aging society with birthrates, lower than the average rates of developed world. Population decline generates many problems such as depopulation in rural areas. One of the measures implemented is to define core areas for maintaining sufficient population density given current and predicted population dynamics. On the other hand, there is a potential for the surroundings of the core areas to be run-down because vacancies generate many problems such as crime, susceptibility to fire, and other negative events. There have been, however, few measures concerning the spatial distribution of parks and open spaces around the core areas. We applied a hedonic approach with a Geographically Weighted Regression (GWR) to the analysis of the relationship between the assessed values of land and geographical information in order to estimate the importance of landscape factors: the spatial continuity of vegetation distributions, public parks, and the local averages of NDVI. It was shown that the number of points where the spatial continuity of vegetation distributions makes positive impacts on nearby land prices is gradually increasing during years 2000 and 2015, while the average of land price continues to fall.

1. INTRODUCTION

Japanese cities are facing a rapidly aging society with birthrates, lower than the average rates of developed world. Population decline generates many problems such as depopulation in rural areas, expansion of lower population density in urban areas, as well as a boost in spending for the maintenance of infrastructure under the pressure of lower revenue. One of the measures implemented is a land use strategy for living space and urban function. In this measure, Residence Attraction Districts and Urban Function Attraction Districts are defined as core areas for maintaining sufficient population density given current and predicted population dynamics. The core areas are expected to play a key role as the fundamental building blocks of future cities in an aging society facing population decline. On the other hand, there is a potential for the surroundings of the core areas to be run-down because vacancies generate many problems such as crime, susceptibility to fire, and other negative events. The government has developed several measures to address the deterioration of the fringe regions of the core areas through the creation of some systems that encourage conversion from vacancies into allotments, citizens' parks, open spaces, and other green spaces. There have been, however, few measures concerning the spatial distribution of parks and open spaces around the core areas. We have developed an advanced analysis method for detecting the spatial continuity of vegetation distributions covering parks and open spaces through the statistical testing of the spatial features of a Normalized Difference Vegetation Index (NDVI) derived from remotely sensed data. The method can achieve the detection of the spatial continuity of vegetation distributions in accordance with the types of land use: a completely urbanized area, a suburban area, a rural area, and a mountainous area. In this study, we identify the economic impacts of the spatial continuity of vegetation distributions through applying several data acquired in 2000, 2008 and 2015. We apply a hedonic approach with a Geographically Weighted Regression (GWR) to the analysis of the relationship between the assessed values of land and geographical information in order to estimate the importance of landscape factors: the spatial continuity of vegetation distributions, public parks, and the local averages of NDVI.

2. METHIDS AND MATERIALS

2.1 Study Area

The whole area of the Osaka prefecture was adopted as the area of interest. This area is located in the Kansai district in the western part of Japan. It covers about 1900 krn^2 and contains 33 cities, 9 towns, and 1 village. There are also watercourses consisting of several rivers and many streams in the area.

2.2 Remote Sensing and Geographical Data

We applied the Landsat series of remote sensing data: ETM+ data from August 2000, TM data from September 2008, and OLS data from September 2015. The data covers the whole area of interest since the observation swath of Landsat sensors is wide enough (185 km). We applied atmospheric corrections based on the MODTRAN for this study. We defined the NDVI calculated from the Landsat data as the proxy of vegetation abundance. Figure 1 shows NDVI derived from the Landsat OLS data as an example. In Figure 1, it is shown that the urbanized areas are bounded on three sides by mountainous areas covered with high NDVI. National Land Numerical Information is used to a hedonic approach including GWR for studying the relationship between the land prices and the attributes of landscape.

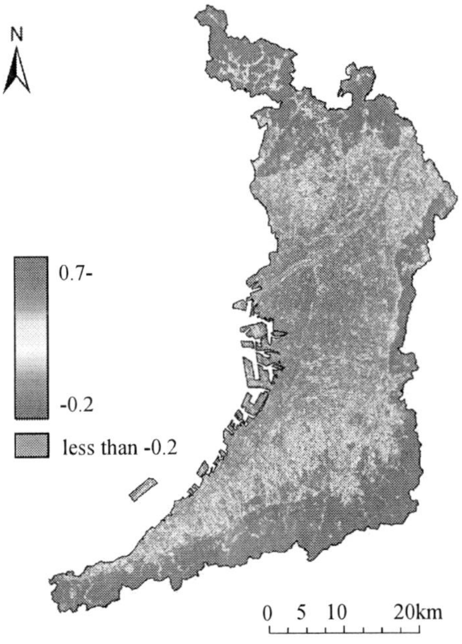

Figure 1. NDVI derived from Landsat OLS data from 2018

2.3 Methods

2.3.1 Generation of SSCs and Ridgelines: The spatial analysis method of vegetation distribution we have developed is composed of a spatial autocorrelation analysis based on G statistics, an overlay analysis, and a hydrological analysis (Getis 1992, Ord 1995, Kumagai 2011). Figure 2 shows procedure of the spatial analysis of vegetation distributions. We overlaid the positive autocorrelation areas generated with the fluctuation of distance *d*: from a widest range to a narrowest range. The area which consists of the multiple layers of the positive autocorrelation area has been called the Spatial Scale of Clumping of vegetated areas (SSC). We also detected the ridgelines from the SSC as the backbones of the high spatial continuity of the vegetated areas by interpreting the SSC as topographic features. The ridgelines play an important role in acting as bridges between the widely dense distribution areas and sparse areas of vegetation, such as the areas (a) and (b) in Figure 2.

Moreover, the application area of the spatial analysis was expanded through the iteration of the procedure with respect to no autocorrelation areas (Kumagai, 2017). Firstly, we applied the spatial analysis method to the area of interest, as mentioned above. The results of this application were renamed SSCl and ridgeline 1. Secondly, we applied the spatial analysis method to the area excluding SSCl, which was initially detected as no spatial autocorrelation area. SSC2 and ridgeline 2 were then generated. Finally, we iterated this procedure to define successive SSCs.

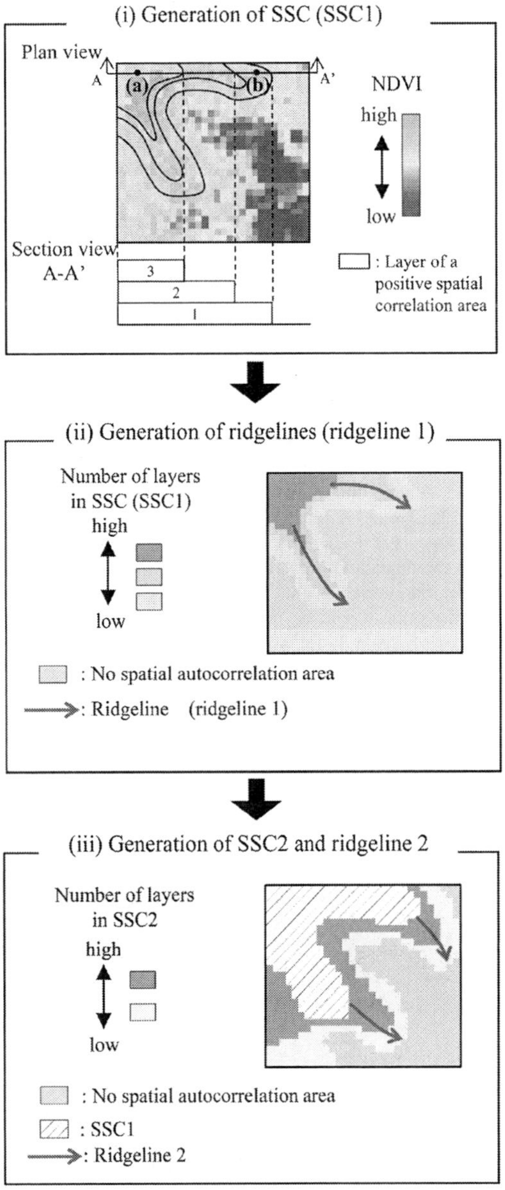

Figure 2. Procedure of the spatial analysis of vegetation distribution

2.3.2 Hedonic method with GWR: To avoid land deterioration around the core areas experiencing population decline in the near future, it is also desirable to be able to conduct a suitable changeover in land use: the substitution of parks and open spaces for vacancies. The changeover should contribute to the maintenance and improvement of the spatial continuity of vegetation distributions. If spatial continuity plays a role in ensuring a rise in value with respect to its surroundings, it will be a good motivating factor to encourage an appropriate changeover of land use. We therefore used a hedonic method with GWR to measure the relationship between the spatial continuity of vegetation distributions and land prices (Brunsdon et al. 1998, Harris et al. 2013, Mulley et al. 2016, Yoo et al. 2016, Hu et al. 2016).

3. RESULTS AND DISSCUSSION

3.1 Results of the Spatial Analysis

Figure 3 indicates the results of the spatial analysis. The study area is divided into 3 SSCs and no autocorrelation area, being common to 2000, 2008, and 2018. The patterns of ridgelines also are slightly different between these results.

3.2 Applicability of GWR

Table 1 shows the variables that we applied to GWR. We adopted the shortest distance from the ridgeline to the point as the factor of spatial continuity of vegetation distributions. The shortest distance to public parks was used as one of the factors of urban facilities. We also calculated the local average of NDVI within 2 km of the point as the local feature of vegetation distributions. These landscape factors are divided into each SSC: SSC1, SSC2, and SSC3 because it appeared that the division could contribute to make an appropriate model to explain land prices (Kumagai 2017).

Before the application of GWR, we examined the relationship between the variables by calculating the VIF. It has been confirmed that there is no multi-collinearity between the variables of Table 1. Initially, we examined the validation of GWR using the statistics of F_1, F_2, and F_3 (Leung et al. 2000). The variables in Table 1 are applied to both OLS and GWR for this examination. The small value of F_1 means that the GWR model has a better goodness of fit than the OLS model, while the large value of F_2 means that the GWR model and the OLS model do not describe the data equally well. F_3 is the test statistic F_k of the spatial differences among the terms of variables. The large value of F_3 means that not all terms are equal. Table 2 shows the results of the validation of GWR. The p values of F_1 and F_2 show less than 0.001 in 2000, 2008, and 2015, respectively. The proportions of significance F_k in F_3 among the three results show more than 70%. Thus, the validity of the GWR model is clarified.

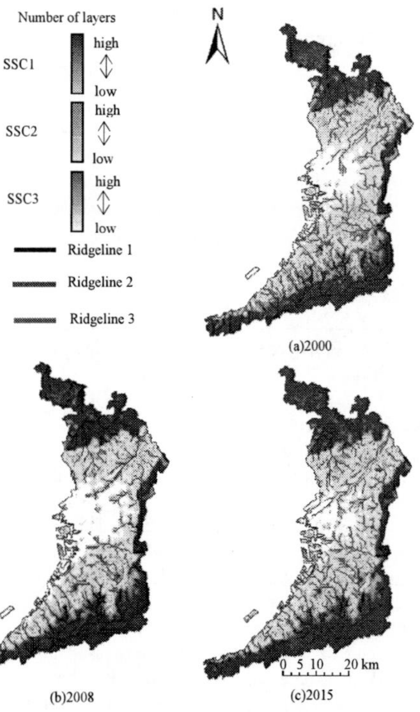

Figure 3. Results of the spatial analysis

	Variables
GI	Acreage [m^2]
	Shortest distance to a railway station [m]
	Building coverage [%]
	Floor area ratio [%]
	Commercial area [dummy]
	Industrial area [dummy]
	Fire protection area [dummy]
	Gas supply [dummy]
	Sewerage [dummy]
Landscape factors	Shortest distance to the ridgeline derived from SSC [m]
	Ridgelines 1, 2, 3, and 4
	Local average of NDVI
	Shortest distance to public parks [m]
	Woods, Small park, and Large park

Table 1. Variables we applied to GWR

Year	F_1	F_2	F_3 (Proportion of F_k(%)) p<0.001	p<0.01
2000	0.735 (p<0.001)	13.11 (p<0.001)	68.0	72.0
2008	0.475 (p<0.001)	9.72 (p<0.001)	88.0	88.0
2015	0.539 (p<0.001)	7.74 (p<0.001)	84.0	84.0

Table 2. Results of the validation of GWR

Year	R^2	AIC	Band width(m)
2000	0.853	210	17088.4
2008	0.896	296	8680.7
2015	0.879	414	9619.3

Table 3. Results of GWR

3.3 Results of GWR

Table 3 displays the results of GWR. R^2 of all the cases are more than 0.85, while AIC and band width seem to depend on the cases. Table 4 shows the proportions of t values of the variable for landscape factors on the basis of the significance level of 5%. The direction (e.g. positive or negative) of an impact on land prices depends on the type of variables. In the variables of shortest distance to public parks, the woods and small parks in SSC2 and SSC3 mostly provide a positive impact on nearby land prices in all years. Especially, the number of t of the small parks in SSC3, showing a positive impact on nearby land prices, gradually increases from 2000 to 2015. NDVI in SSC1 and SSC2 nearly indicates a negative impact in all cases, while the number of t showing a negative impact in SSC3 increases by degrees during these periods. It is also indicated that there are few points in all SSCs, where NDVI shows a positive impact, in all years. On the other hands, in shortest distance to the ridgeline derivied from SSC, the number of t values of the ridgeline 3 in SSC3, making positive impacts on

This contribution has been peer-reviewed.
https://doi.org/10.5194/isprs-archives-XLII-3-W7-41-2019 | © Authors 2019. CC BY 4.0 License.

	t > 1.960 (p(>t)=0.025)	1.960 >= t >= -1.960	-1.960 > t (p(<t)=0.025)
Shortest distance to the ridgeline derived from SSC [m]			
Ridgeline 1 in SSC1	90.69	9.31	0.00
Ridgeline 2 in SSC2	0.00	91.95	8.05
Ridgeline 3 in SSC3	0.00	99.85	0.15
Local average of NDVI			
NDVI in SSC1	0.00	0.00	100.00
NDVI in SSC2	0.00	33.62	66.38
NDVI in SSC3	0.00	100.00	0.00
Shortest distance to public parks [m]			
Woods in SSC1	53.50	20.18	26.32
Woods in SSC2	0.00	17.41	82.59
Woods in SSC3	0.00	2.47	97.53
Small park in SSC1	0.00	0.96	99.04
Small park in SSC2	0.00	6.24	93.76
Small park in SSC3	0.00	59.03	40.97
Large park in SSC1	30.60	69.25	0.15
Large park in SSC2	72.32	20.38	7.30
Large park in SSC3	78.86	21.14	0.00

(a)2000

	t > 1.960 (p(>t)=0.025)	1.960 >= t >= -1.960	-1.960 > t (p(<t)=0.025)
Shortest distance to the ridgeline derived from SSC [m]			
Ridgeline 1 in SSC1	47.72	38.10	14.18
Ridgeline 2 in SSC2	13.67	77.73	8.60
Ridgeline 3 in SSC3	0.00	79.72	20.28
Local average of NDVI			
NDVI in SSC1	3.69	32.87	63.44
NDVI in SSC2	0.00	51.46	48.54
NDVI in SSC3	0.00	26.01	73.99
Shortest distance to public parks [m]			
Woods in SSC1	15.46	48.23	36.30
Woods in SSC2	0.00	74.81	25.19
Woods in SSC3	0.56	54.58	44.85
Small park in SSC1	0.00	2.82	97.18
Small park in SSC2	0.00	15.21	84.79
Small park in SSC3	0.00	56.89	43.11
Large park in SSC1	60.68	36.82	2.51
Large park in SSC2	35.54	45.21	19.25
Large park in SSC3	54.02	45.98	0.00

(b)2008

	t > 1.960 (p(>t)=0.025)	1.960 >= t >= -1.960	-1.960 > t (p(<t)=0.025)
Shortest distance to the ridgeline derived from SSC [m]			
Ridgeline 1 in SSC1	0.00	90.79	9.21
Ridgeline 2 in SSC2	0.00	100.00	0.00
Ridgeline 3 in SSC3	4.02	15.96	80.01
Local average of NDVI			
NDVI in SSC1	0.00	33.08	66.92
NDVI in SSC2	0.00	30.08	69.92
NDVI in SSC3	0.00	21.83	78.17
Shortest distance to public parks [m]			
Woods in SSC1	10.30	59.35	30.35
Woods in SSC2	0.00	16.71	83.29
Woods in SSC3	0.00	47.61	52.39
Small park in SSC1	5.53	86.43	8.05
Small park in SSC2	0.00	76.33	23.67
Small park in SSC3	0.00	20.94	79.06
Large park in SSC1	0.00	71.83	28.17
Large park in SSC2	39.22	51.57	9.21
Large park in SSC3	66.58	33.42	0.00

(%)

(c)2015

Table 4. Proportions of t values of the variable in landscape factors are indicated on the basis of the significance level of 5%

nearby land prices, is gradually increasing during years 2000 and 2015.

3.4 Comparison between Land Prices and Results of GWR

Figure 4 shows the fluctuation of land prices in SSC3 during years 2000 and 2015. In Japan, economic bubble occurred from the mid-1980s to the early 1990. After this periods, land prices generally had declined for about 20 years on the basis of economic slump. We can see the distributions of land prices being on the decrease during years 2000 and 2015 in Figure 4.

Figure 5 displays the distributions of t values in SSC3. Warm

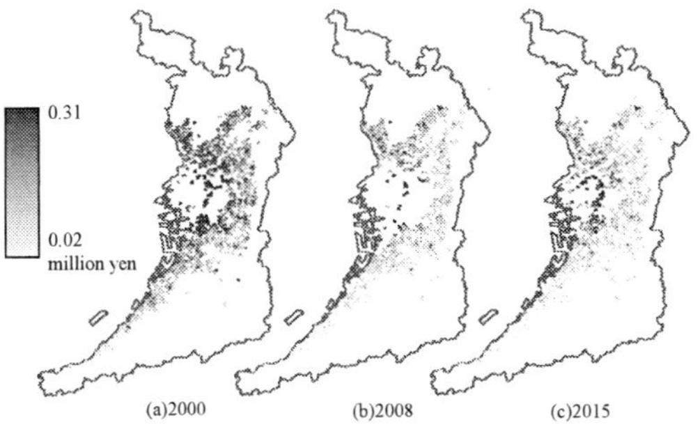

Figure 4. Distribution of land prices in SSC3

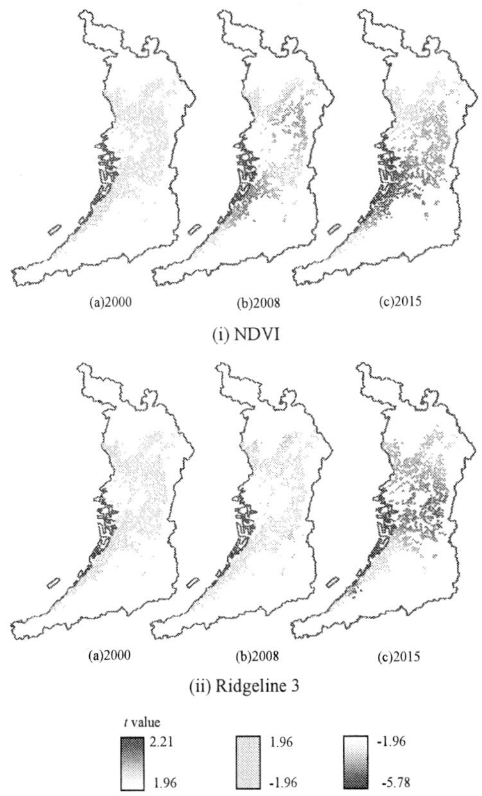

Figure 5. Spatial distribution of t values in SSC3

colour points mean that the t value is above the significance level of 5%, while cold colour points mean that the t value is below the significance level of 5%. It is apparent that the number of t values of the shortest distance to ridgeline 3, being below the significance level of 5%, increases as the land prices decrease during years 2000 and 2015. Likewise, the number of t value of NDVI, being below the significance level of 5%, increases. The local averages of NDVI in 2015 seem to perform a general role in this model: usually, the land prices of urbanized areas including very few green spaces are relatively high, while the land prices of rural areas surrounded by large green spaces are relatively low. The local average of NDVI, however, does not make a positive / negative impact on nearby land prices in the case where land prices are relatively high. The shortest distance of ridgeline 3 also does not make a positive / negative impact on nearby land prices in 2000. The number of points where the shortest distance of the ridgeline 3 makes a positive impact on nearby land prices is increasing as land prices are decreasing during years 2000 and 2015. In other words, the spatial continuity of vegetation distributions seems to play a more important role in contributing to a substantial extent to predict the land prices of suburban areas around the core areas as land prices decrease.

4. CONCLUSOIN

In this study, we applied the multitemporal analysis of the relationship between assessed values of land and landscape factors by using geographical information data and remotely sensed data acquired in 200, 2008, and 2015. On the basis of the application of hedonic method with GWR, the landscape factors derived from NDVI hardly have a positive / negative impact on nearby land prices in case where the average of land prices are relatively high. It was shown that the number of points where the spatial continuity of vegetation distributions makes positive impacts on nearby land prices is gradually increasing during years 2000 and 2015, while the average of land price continues to fall.

REFERENCES

Brunsdon, C., Fotheringham, A. S., and Chariton, M., 1998. Geographically weighted regression-modelling spatial non-stationarity. *The Statistician*, 3(47), pp.431-443.

Getis, A., and Ord, J. K., 1992. The analysis of spatial association by use of distance Statistics. *Geographical Analysis*, 24(3), pp.189-206.

Harris, R., Dong, G., and Zhang, W., 2013. Using Contextualized geographically weighted regression to model the spatial heterogeneity of land prices in Beijing, China. *Transactions in GIS*, 17(6), pp.901- 919.

Hu, S., Yang, S., Li, W., Zhang, C., and Xu, F., 2016. Spatially non-stationary relationships between urban residential land price and impact factors in Wuhan city, China. *Applied Geography*, 68, pp.48- 56.

Kurnagai, K., 2011. Verification of the analysis method for extracting the spatial continuity of the vegetation distribution on a regional scale. *Computers, Environment and Urban Systems*, 35, pp.399-407.

Kumagai, K., Uematsu, H., and Matsuda, Y., 2017. Advanced Spatial Analysis for Vegetation Distributions Aimed at Introducing Smarter City Shrinkage. *Planning Support Science for Smarter Urban Futures, Lecture Notes in Geoinformation and Cartography*. Springer International Publishing, pp.469-489.

Leung, Y., Mei, C., and Zhang, W. X., 2000. Statistical tests for spatial nonstationarity based on the geographically weighted regression model. *Environment and Planning A*, 2000(32), pp.9-32.

Mulley, C., Ma, L., Clifton, G., Yen, B., and Burke, M., 2016. Residential property value impacts of proximity to transport infrastructure: An investigation of bus rapid transit and heavy rail networks in Brisbane, Australia. *Journal of Transport Geography*, 54, pp.41- 52.

Ord, J. K., and Getis, A., 1995. Local spatial autocorrelation statistics: Distributional issues and an application. *Geographical Analysis*, 27(4), pp.286-306.

Yoo, J., and Ready, R., 2016. The impact of agricultural conservation easement on nearby house prices: Incorporating spatial autocorrelation and spatial heterogeneity. *Journal of Forest Economics*, 25, pp.78-79.

GROUND PENETRATING RADAR DATA INTERPRETATION USING ELECTROMAGNETIC FIELD ANALYSIS FOR SEA ICE THICKNESS MEASUREMENT

M. Matsumoto [1], M. Yoshimura [2], K. Naoki [3], K. Cho [3], H. Wakabayashi [4]

[1] PASCO CORPORATION, 2-8-10, Higashiyama, Meguro-ku, Tokyo, Japan – moatsy7605@pasco.co.jp
[2]Center for Spatial Information Science, The University of Tokyo, Meguro-ku, Tokyo, Japan
[3] Tokai University, 2-28-4, Tomigaya, Shibuya-ku, Tokyo, Japan
[4] Nihon University, 1 Nakagawara, Tokusada, Tamura, Koriyama, Japan

Commission III, WG III/2

KEY WORDS: Sea ice thickness measurement, Ground truth, Ground Penetrating Radar, Brackish lake, Finite-Difference Time-Domain method, Dielectric constant, Multiple reflection

ABSTRACT:

Observation of sea ice thickness by remote sensing is one of key issues to understand an effect of global warming. However, ground truth must be necessary to discuss this kind of approach. Although there are several methods to acquire ice thickness, Ground Penetrating Radar (GPR) can be good solution because it can discriminate snow-ice and ice-sea water interface thanks to comparative higher spatial resolution than the other methods. In this paper, we carried out GPR measurement in brackish lake and an electromagnetic field analysis in order to interpret the GPR data. The simulation model was assumed considering the actual snow and ice thickness acquired in field measurement. From the simulation results, although it seems difficult to identify the reflection at snow and ice interface due to a thin layer thickness and a low dielectric constant, snow and ice thickness may be estimated by using multiple reflection components.

1. INTRODUCTION

Sea ice thickness is one of important parameters to understand an effect of global warming. The remote sensing approach such as Synthetic Aperture Radar (SAR) has possibility to estimate ice thickness (Wakabayashi et al., 2004a). However, still ground truth must be necessary to discuss the remote sensing approach.

There are several methods to obtain the ground truth of ice and snow thickness. Simplest but most correct way is making a hole and measuring the thickness directly by a ruler. However this method is destructive, time consuming and gives only point information. Although, there are other methods in order to cover larger area such as electromagnetic induction, sonar and so on, ground penetrating radar (GPR) can be good solution because it can discriminate snow-ice and ice-sea water interface thanks to comparative higher spatial resolution than the other methods (Liu et al., 2014a). GPR is a tool in order to image a subsurface structure (shallower than several meters) such as buried object, layer structure and so on by using a penetration and reflection of microwave. The GPR can also be used in ice thickness measurement (Tebaldini et al., 2016a).

On Feb. 2018, we carried out GPR measurement in Lake Saroma, which is a brackish lake, for relatively large area (200m by 300m, approximately) aiming to obtain grand truth for remote sensing data (Matsumoto et al., 2018). Although several Bscan images could be obtained, it was difficult to identify where reflection signals come from. In this paper, we carried out an electromagnetic filed analysis in order to interpret Bscan image acquired in field measurement. Firstly, methodology for ice thickness estimation and field experiment in Lake Saroma will be explained. Considering the result of the direct measurement, the simulation model was assumed and the electromagnetic field analysis was implemented. The interpretation of simulated result will be presented.

2. METHODOROGY

Figure 1 shows skematic view of ice thickness measurement by GPR. A three layer structure, which consists of snow, ice and water, is assumed. GPR transmits electromagnetic wave into subsurface and received reflected wave caused by snow-ice and ice-water interface. GPR records amplitude and arrival time of reflection signal. Therefore if we identify the each reflection signal, layer thickness H can be determined by

$$H = \frac{1}{2} v \Delta t \qquad (1)$$

where Δt = time difference of reflection signal
v = propagation velocity in a medium and given by

$$v = \frac{c}{\sqrt{\varepsilon_r}} \qquad (2)$$

where c = speed of electromagnetic wave in vacuum
ε_r = dielectric constant of a medium

Note that the propagation velocity, which is determined by dielectric constant of material, is still unknown.

Our methodology to estimate snow and ice thickness is as following. Firstly, we carry out GPR survey and acquire cross section image, which is called Bscan image, by collecting a reflection signal along survey line. The direct measurement on the survey line should be conducted just after GPR survey at several points for calibration and validation. If the reflection signal from interface can be identified in Bscan image, time difference of the signal also can be determined. By considering the actual thickness of snow and ice, the propagation velocity at which direct measurement carried out can be determined. Therefore snow and ice thickness along survey line can be estimated by assuming that the propagation velocity is constant. In this method, the important points are 1) Whether the reflection signal can be identified, and 2) How the arrival time of the signal can be determined.

An electromagnetic field analysis was conducted in order to confirm these points. The gprMax which is a free numerical GPR simulator based on FDTD (Finite-Difference Time-Domain) method (Warren et al., 2016a) was used for this purpose. We assumed a three layer model which consists of snow, ice and sea water, having layer thickness of 20cm, 20cm and 40cm, respectively. The dielectric constant of them is assumed as 2, 3 and 81, and the conductivity is assumed as 10^{-5}S/m, 10^{-3}S/m, and 100S/m, respectively by referring to a list of electric conductivity (Daniels, 2004). A case of dry snow was assumed in this simulation. This case may be more difficult because travel time of snow layer is shorter due to lower dielectric constant resulting in a difficulty to discriminate the reflection from snow-ice interface and direct antenna coupling. A 600MHz GPR system was modelled by using a first derivative of Gaussian pulse with central frequency of 600MHz as a feeding source. Figure 2 shows simulated signal. Absolute value of analytic signal representation is used in order to detect a single peak of the reflection signal. The reflection signals at t_1, t_2 and t_3 correspond to Path 1, Path 2 and Path 3 in Figure 1, respectively.

By calculating the time difference of peak of reflection signal, the thickness of snow and ice were estimated as 18.95cm and 19.23cm, respectively. The results were less than 5% error and almost agreed with actual thickness in the simulation model. Therefore this method can be work if we identify the peak of reflection signal from layer interface.

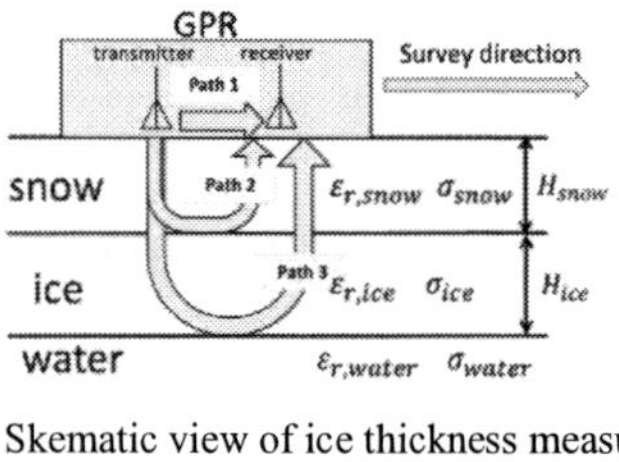

Figure 1. Skematic view of ice thickness measurement.

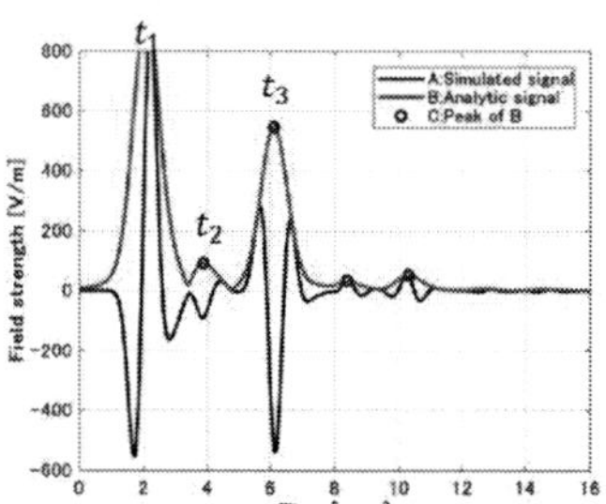

Figure 2. Reflected wave simulated by gprMax.

3. FIELD EXPERIMENT

3.1 Test Field

Field experiment was carried out in Lake Saroma, north east of Hokkaido, Japan, on 18th to 22 of February 2018. Lake Saroma is facing to Sea of Okhotsk and connected by two inlets resulting in brackish lake. Figure 3 (a) shows Satellite image around the test field located in west side of the lake. The area indicated by an arrow can be discriminated by the crack as shown in Figure 3 (b). Figure 3 (c) shows location of GPR survey lines in the right side of the crack area indicated by rectangular in Figure 3 (a). GPR survey was conducted at five locations in the area.

3.2 Measurement

Figure 4 shows a GPR system used in the experiment. The radar unit is "RIS Hi-Mod", product of IDS GeoRadar s.r.l. The Bscan image of 600MHz was obtained by collecting the reflection signal at every 3cm using hip chain encoder along the scan direction. A direct measurement along survey line was also conducted in No.1, No.2 and No.4 with interval of 3m as shown in Figure 3 (c). We drilled and measured thickness of snow and ice by a ruler directly.

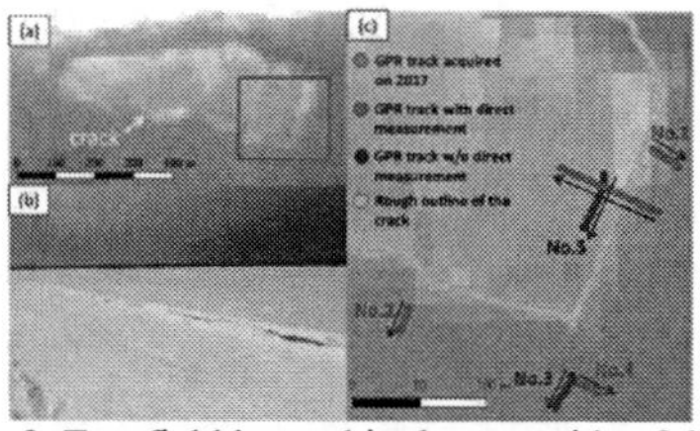

Figure 3. Test field located in the west side of the lake.
(a) Sattellite image taken by Sentinel-2 (Jan. 18th, 2018).
(b) Crack arround the area. (c) GPR survey line arround the rectanguler in (a). The arrows indicate scan direction.

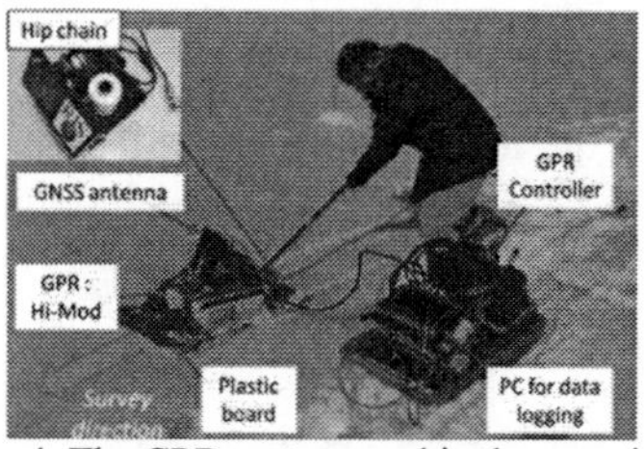

Figure 4. The GPR system used in the experiment.

3.3 Result

According to the direct measurement, the overall average and standard deviation of snow were 10.1cm and 2.2cm, and those of ice were 26.0cm and 3.0cm, respectively. We also observed the condition between snow and ice layer. As for No.1, there was a mixture of snow and water above ice layer at distance of 9m to 18m. As for No.2 and No.4, we found that snow just above ice layer is bit wet condition at all points.

Figure 5 shows GPR Bscan images acquired in survey line No.1, No.2 and No.4, respectively. Horizontal axis indicates GPR position acquired by hip chain encoder. Vertical axis indicates arrival time. The layer like structure can be seen in part of Figure 5 (a) and (b), and whole of (c) as indicated by arrows. Although these signal may be a reflection which is related to snow-ice and ice-water interface, it is not obvious where the reflections come from. Besides, since the average snow thickness is thinner than the simulation model in chapter 2, the method to estimate thickness may be different to apply. Therefore the simulation model must be adjusted by considering the actual thickness acquired in the direct measurement in order to interpret the GPR data.

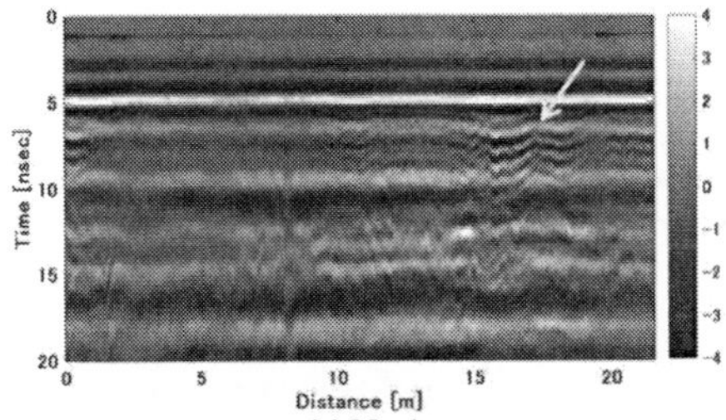

(a) No.1

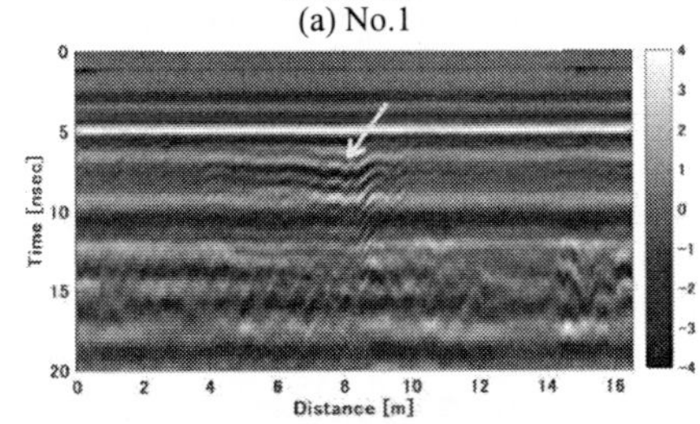

(b) No.2

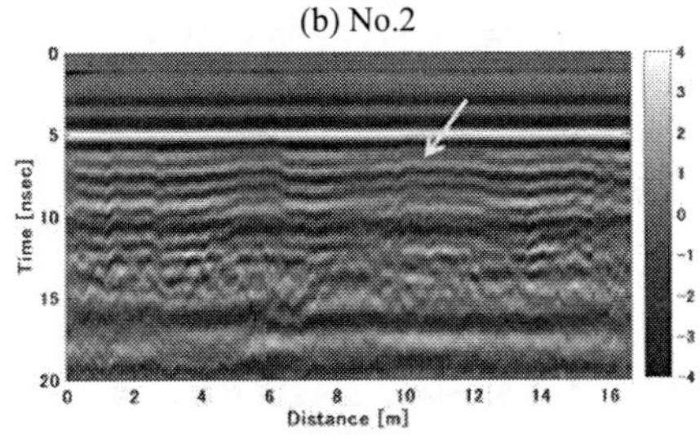

(c) No.4

Figure 5. Bscan image obtained from 600MHz antenna.

4. NUMERICAL ANALYSIS

Aiming to interpret the Bscan images, we carried out an electromagnetic field analysis. We assumed layer thickness of snow, ice and sea water is 10cm, 26cm and 44cm, respectively, considering the result of direct measurement. The dielectric constant and the conductivity are same value mentioned in chapter 2. We call this model as Model 1. Besides, we assumed other two models, Model 2 and Model 3 which have different layer thickness in order to identify reflection signal from layer interface. Figure 6 shows these simulation models. Compared to Model 1, Model 2 has same thickness of snow layer but 10cm thicker ice, on the other hand, Model 3 has 10cm thicker snow but same thickness of ice.

Figure 7 shows path of reflection signal in simulated models. We assumed six reflection paths such as P_0 to P_5 as shown in this figure. Three interfaces such as air-snow, snow-ice and ice-sea water are represented as B_1, B_2 and B_3, respectively. Two-way travel time in the mediums is represented as T_1, T_2 and T_3, respectively. P_0 and T_0 indicate antenna direct coupling and its travel time. P_1, P_2 and P_3 are paths reflected at B_1, B_2 and B_3, respectively. P_4 and P_5 indicate a path of multiple reflections. As for P_4, firstly, the electromagnetic wave is reflected at B_3. After that the electromagnetic wave is reflected at B_2 and B_3 then back to the antenna. As for P_5, the electromagnetic wave is reflected at B_3, B_1 and B_3 then back to the receiver. The reflection coefficient of layer interface of medium1 and medium 2 can be calculated by following formula.

$$\Gamma = \frac{\sqrt{\varepsilon_{r1}}-\sqrt{\varepsilon_{r2}}}{\sqrt{\varepsilon_{r1}}+\sqrt{\varepsilon_{r2}}} \quad (3)$$

According to (3), absolute values of reflection coefficients of air-snow, snow-ice and ice-sea water interface are 0.17, 0.10 and 0.68, respectively.

Figure 8 shows simulation results of assumed models. Analytic signal representation is used in order to discuss the arrival time of the peak signal. The four main reflections were observed in Figure 8 (a). Although the peak signal which may come from snow-ice interface can be observed in Figure 2, this peak cannot be observed in Figure7 (a) due to thinner snow thickness. Therefore it is not simple to calculate layer thickness compared to the case of Figure 2.

Let us focus on an analysis of Figure 8 (a), i.e., Model 1. First peak can be mainly a direct antenna coupling because the direct antenna coupling has much larger amplitude than the other reflection signals. The reflection from air-snow interface is much weaker and distance between antenna and snow is too close, the reflection path P_1 cannot be identified. The reflection path P_2 seems to be also included at the first peak due to thin snow thickness. If snow thickness is thicker as Model 3, the reflection path P_2 can be observed as shown in Figure 8 (c). Therefore we concluded that reflection signal at snow-ice interface may not be observed in measurement data due to thin snow thickness. Summarizing the interpretation above, the first peak includes the reflection path P_0, P_1 and P_2.

Second peak of signal which has second strongest signal amplitude can be reflection path P_3. Observing the result of Model 2 and Model 3 which are the deeper water level from top of snow as shown in Figure 8 (b) and (c), arrival time of this peak are later than the case of Model 1. Besides, absolute value of reflection coefficient at B_3 is larger than the other interface. This can be the reason why the signal amplitude is stronger than the others. Therefore we concluded that second peak of the signal comes from ice-sea water interface, i.e., reflection path P_3.

Third and fourth peak of signal can be considered that they correspond to reflection path P_4 and P_5, i.e., multiple reflections. If snow is thicker as Model 3, time difference of P_4 and P_5 is larger as shown in Figure 8 (c). The time difference of them is 0.85nsec for Model 1, 0.90nsec for Model 2 and 1.93nsec for Model 3, respectively. If ice is thicker as Model 2, time difference of P_3 and P_4 is larger as shown in Figure 8 (b). The time difference of them is 3.03nsec for Model 1, 4.13nsec for Model 2 and 2.96nsec for Model 3, respectively. From this observation, third and fourth peak of signal can be considered as reflection path P_4 and P_5.

The travel time of each reflection path is shown in Figure 7. From this figure, the snow thickness can be estimated by the time difference of P_4 and P_5. The ice thickness can be estimated by that of P_3 and P_4. As for Model 1, the snow and ice thickness was estimated as 9.0cm and 26.22cm, respectively. The errors of them are 10% and 0.85%, respectively. Even though the reflection at snow-ice cannot be identified, we confirmed that the snow and ice thickness can be estimated using time difference of second, third and fourth peak of signal. Therefore we concluded that multiple reflections are useful information for the thickness estimation if the snow layer is thin and the reflection from snow-ice interface cannot be identified directly. The comparison of simulated data and measured GPR data will be a future work.

https://doi.org/10.5194/isprs-archives-XLII-3-W7-47-2019

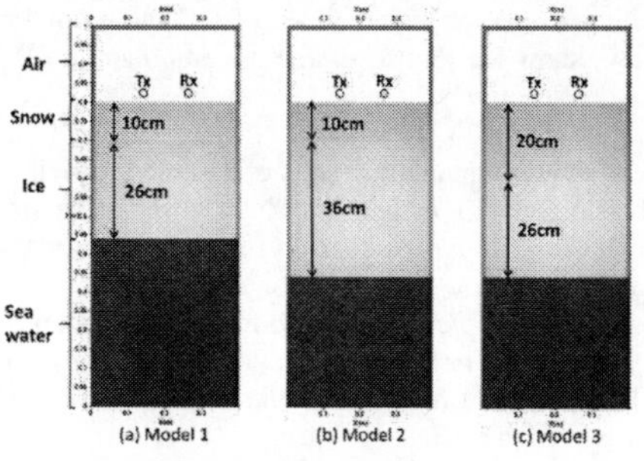

Figure 6. Simulation model

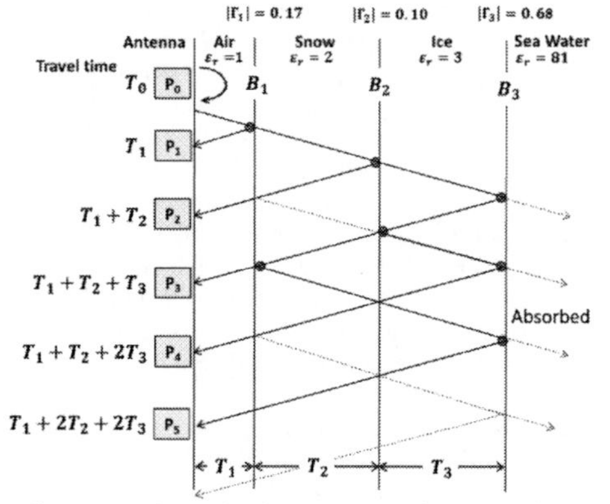

Figure 7. Schematic diagram of multiple reflection

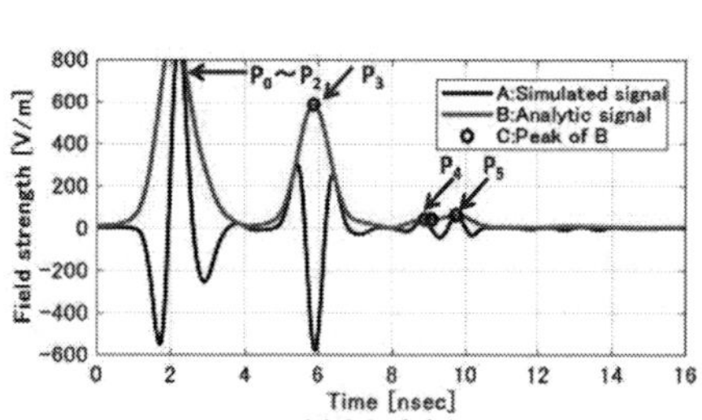

(a) Model 1

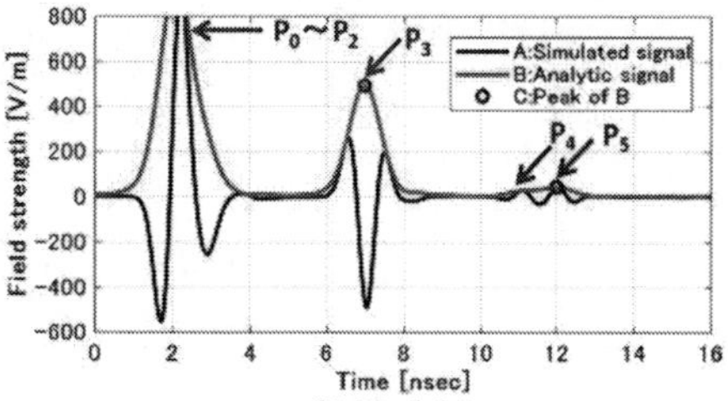

(b) Model 2

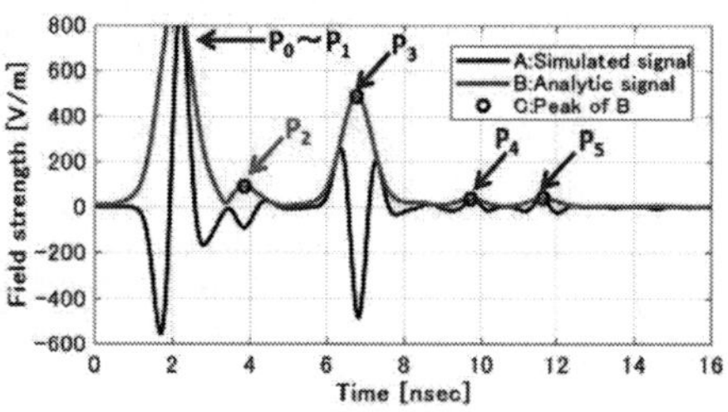

(c) Model 3

Figure 8. Simulation results.

5. CONCLUSION

In this paper, an electromagnetic field analysis was carried out in order to interpret GPR data which was acquired as a ground truth of snow and ice thickness for remote sensing data. Simulation model was assumed by considering the result of direct measurement. From the simulation results, although it seems difficult to identify the reflection at snow and ice interface due to thin layer thickness and low dielectric constant, snow and ice thickness may be estimated by using multiple reflection components. The comparison of simulated data and measured GPR data will be a future work.

ACKNOWLEDGEMENTS

The authors acknowledge Mr. T. Sato for his support in the field measurement.

REFERENCES

Daniels, D.J. (Ed.), 2004. *Ground Penetrating Radar 2nd Edition*. The Institution of Engineering and Technology, London.

Liu, H., Takahashi, K., Sato, M., 2014a. Measurement of Dielectric Permittivity and Thickness of Snow and Ice on a Brackish Lagoon Using GPR. *IEEE Journal of Selected Topics in Applied Earth Observations and Remote sensing*, Vol.7, No.3. pp. 820-827.

Matsumoto, M., Yoshimura, M., Naoki, K., Cho, K., Wakabayashi, H., 2018. SEA ICE THICKNESS MEASUREMENT BY GROUND PENETRATING RADAR FOR GROUND TRUTH OF MICROWAVE REMOTE SENSING DATA. *The International Archives of the Photogrammetry, Remote Sensing and Spatial Information Sciences,* Vol. XLII-3, 2018, pp. 1259-1262.

Tebaldini, .S, Nagler, T., Rott, H., Heilig, A. 2016a. Imaging the Internal Structure of an Alpine Glacier via L-Band Airborne SAR Tomography. *IEEE Transactions on Geoscience and Remote sensing*, Vol.54, No.12. pp. 7197-7209

Wakabayashi, H., Matsuoka, T., Nakamura, K., Nishio, F., 2004a. Polarimetric Characteristics of Sea Ice in the Sea of Okhotsk Observed by Airborne L-Band SAR. *IEEE Transaction on Geoscience and Remote Sensing*, Vol.42, No.11. pp. 2412-2425.

Warren, C., Giannopoulos, A., Giannakis, I., 2016a. gprMax: open source software to simulate electromagnetic wave propagation for Ground Penetrating Radar, *Computer Physics Communications,* Vol.209, p.163-170.

This contribution has been peer-reviewed.
https://doi.org/10.5194/isprs-archives-XLII-3-W7-47-2019

USING REMOTE SENSING DATA FOR CALCULATING THE COASTAL EROSION IN SOUTHERN THAILAND

W. Pantanahiran [1]

[1] Dept. of Geography, Faculty of Social Sciences, Srinakharinwirot University, Bangkok, Thailand, - wichaip@swu.ac.th, pwichai@yahoo.com

Commission WG III/10

KEY WORDS: Coastal Erosion, Geographic Information System, Remote Sensing, IKONOS, Quick Bird, Worldview-2, Pleiades

ABSTRACT:

There has been long-term observation of coastal erosion in Koh Kho Khao , Ban Nam Khem, Phang Nga province, in Thailand, which was affected by a tsunami on December 26, 2004. The disaster, as is well known, caused the loss of lives and property. This area is recognized as one of the best tourist areas in Thailand. The objective of the research was to identify the coastal changes to the island, Koh Kho Khao. The Geographic Information System and Remote Sensing were used. Five- time periods were used, in which aerial photographs and satellite images were taken, with the aerial photographs taken in February, 2002. IKONOS images were taken on December 29, 2004, and Quick Bird images were dated the 23rd of February, 2009. Worldview-2 images were dated the 6th of December, 2012, while the Pleiades images were dated January 8, 2016. The coastlines were compared using the overlay technique. Coastal erosion and coastal deposition during consecutive years were calculated. The results showed that the tsunami in 2004 caused coastal erosion in the area, as coastal changes during those years were found. Additionally, natural adaptation was found after 14 years at the middle and upper parts of the island. Severe coastal erosion of the lower part of the island has been continuously found, with an erosion rate between 2002 and 2004 (2-year period), 2004 and 2009 (5-year period), 2009 and 2012 (3-year period), and 2012 and 2016 (4-year period) of 22.44, 9.96, 19.63, and 12.34 meters per year respectively. In addition the erosion rate between 2002 and 2016 (14-year period) was 100.97 meters per year. It was also found that the seawall was the main factor in the coastal erosion in the lower part of the island because it was recognized that the coastline was sharply cut along the seawall. It is recommended that the lower part of the island be declared a special observation area in order to prevent further coastal erosion.

1. INTRODUCTION

Coastal areas are important for economic, social, and environmental reasons. Various uses of coastal zones, including tourism and recreation activities, fisheries and aquaculture, community activities, ports, industries, and water transportation, can cause changes in coastal areas. Serious socio-economic problems may arise with erosion and the loss of beaches (Clark, 1996). Coastal change means changes in physical characteristics of the coast due to the erosion of coastal areas and the accumulation of sediment (Sinsakul et al., 2003). Climate change can also aggravate the deterioration of natural resources and the environment. Nowadays, the erosion of the coast is more frequent and more severe.

Tsunami waves are waves that originate in the deep sea which often appear after a large earthquake underwater and intensified by volcanic eruptions, landslides, land subsidence, or when large meteorites fall into the sea. A tsunami can destroy coastal areas, causing loss of life and property. Tsunami waves cannot be predicted a long time in advance but there can be short-term warnings of them. On the 26th of December, 2004, a 9.0 scale earthquake occurred in the northwest part of Sumatra, Indonesia that caused a tsunami, one of the most serious geological events in world history. This event wreaked havoc for many countries in the Indian Ocean and the Andaman Sea areas, including Indonesia, Malaysia, Sri Lanka, Myanmar, Bangladesh, India, the Maldives, Somalia, Kenya, and Thailand. This event caused more than 250,000 deaths, 500,000 injuries, and more than 2 million people were left homeless.

The coastal provinces in the Andaman areas of Thailand were affected, including Krabi, Phuket, Trang, Phang Nga, Ranong, and Satun. More than 5,219 local people and foreigners died, 3,498 people disappeared, and 8,457 were injured. In addition, a great amount of the natural resources in the areas and the tourism industry were affected (Department of Mineral Resources, 2006). The impact of the tsunami caused severe and extensive coastal changes in Koh Kho Khao (Pantanahiran, 2014). Natural disasters in Thailand have tended to be more frequent and more violent in recent years due to climate changes, and this has affected the economic development of the country as well as the way of life of the people.

This research studied the physical changes in the coastal areas mentioned above by using aerial photographs and satellite images. The coastlines were compared in order to calculate the erosion and deposition in the areas. A Geographic Information Systems (GIS) was used to analyse the data. A GIS refers to a collection of computer hardware, software, and geographic data for capturing, storing, updating, manipulating, analysing, and displaying all forms of geographically-referenced information (Kennedy, 2001).

Remote sensing on the other hand is the acquisition of information concerning an object or phenomenon without making physical contact with the object and thus is in contrast to on-site observation, especially the earth. Remote sensing is used in numerous fields, including geography, land surveying, and most of the earth science disciplines.

The objective of the research was to identify the coastal changes to the island, Koh Kho Khao in Takua Pa District, Phang Nga Province. The study area was divided into 3 parts: the upper area, the middle area, and the lower area (Figure 1).

1.1 Description of study area

The study area was located in Koh Kho Khoa (island), Ban Nam Khem sub-district, Takua Pa district, Phang Nga province, in the Andaman sea in southwest Thailand. The area is nearly flat, and the elevation is about two meters above mean sea level (MSL). The area of the island is approximately 6,328.33 ha (Figure 1). The land use has been classified as coastal zone (22.32 ha), upland crops (271.41 ha), orchards (605.78 ha), paddy rice (33.68 ha), upland forests (1,968.14 ha), shrubs (697.76), mangrove forests (2,341.19 ha), wetlands (199.18 ha), and unclassified (83.26 ha). Additionally, the climate has been classified as tropical monsoon. Takua Pa District is fully influenced by the northeast monsoons from January to April during the summer season and the southwest monsoons from May to December during the rainy season. The island is also well-known as a touristic area (Figure 1).

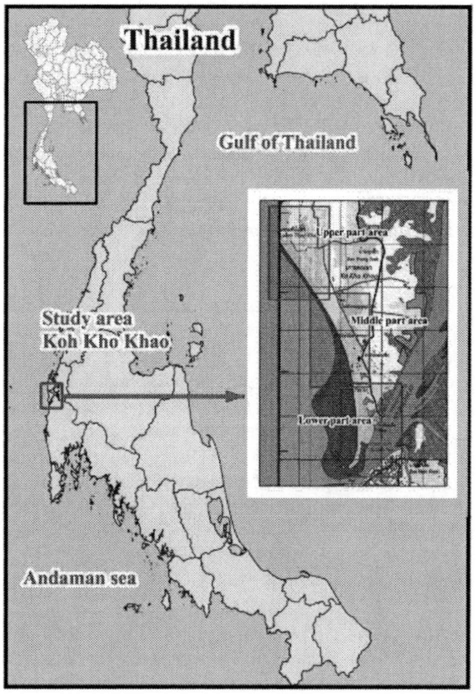

Figure 1. Study area

2. MATERIAL & METHODS

2.1 Data input and analysis

Remote sensing technology (RS) and the geographic information system were used as the analytical tools, and the ArcGIS software was used. The Universal Transverse Mercator (UTM) coordinate system was used as a standard coordinate system. The coastlines in the study refer to the beach ridges or slope changes that could be detected on the digital orthophotographs. The erosion and deposition/accretion of the coastal zone were detected by using the overlay technique between two sets of coastal lines (Pantanahiran, Weesakul, and Thaicharoen, 2008; Pantanahiran, 2014)

The remotely-sensed data in the study area were selected, including data before and after the tsunami on December 26th, 2004. The data were collected over time. The aerial photographs (scale 1:25000) taken in February 2002 represent the coastal area before the Tsunami. Then, satellite images after the tsunami were collected (Table 1). The IKONOS imagery (4-meter resolution) was taken on December 29, 2004. The QuickBird imagery (1-meter resolution) was taken on February 23, 2009, and the Worldview-2 imagery (1.8-meter resolution) was taken on December 6, 2012. The Pleiades imagery (0.5-meter resolution) was taken on the 8th of January, 2016, (Table 1). Those data were registered by using the appropriate geographic points of reference (ground control point). As such, the data could be completely overlaid and represent the same location.

Events	Type of data	Scale/ Resolution (m)	Date
Before tsunami	Aerial photographs	1:25000	February 2002
Three days after tsunami	IKONOS	4	29 December 2004
Five years after tsunami	QuickBird	1	23 February 2009
Eight years after tsunami	Worldview-2	1.8	6 December 2012
Twelve years after tsunami	Pleiades	0.5	8 January 2016

Table 1 Data used for coastal erosion.

The coastlines were digitized from the aerial photographs and satellite images representing the situation of the coastal area. The erosion and deposition of the coastal area were detected by using the overlay technique between two coastlines at different time periods. The comparison of the erosion and deposition was calculated using the coastlines between 2002 and 2004, the coastlines between 2004 and 20122, the coastlines between the years 20122 and 2012, the coastlines between the years 2012 and 2016, and the coastlines between 2002 and 2016.

Further, the beach erosion or deposition areas were calculated. The beach erosion indicates the retreat of the coastlines landward, and the deposition or accretion indicates the seaward deposition of the sand. It is reasonable to assume that the area means the product of its length and its width, and the erosion (width) may be calculated by the area divided by the erosion length along the shoreline. Then, the erosion and the rate of erosion were calculated (Pantanahiran, Weesakul, and Thaicharoen, 2008) and (ONEP 2003) using the following equations:

$$x = \frac{a}{l} \quad (1)$$

$$y = \frac{x}{t} \quad (2)$$

where x = beach erosion (m)
a = erosion area (m2)
l = distance of erosion along the coastline (m)
y = average erosion rate (m/y)
t = *comparing years (years).*

https://doi.org/10.5194/isprs-archives-XLII-3-W7-51-2019

The calculated result of the erosion or deposition rate was classified into three main groups: the erosion coast, the deposition coast, and the stable coast (Pantanahiran, 2014) and (Sinsakul et al., 2003). The severe erosion means the rates of change are greater than -5 meters per year (m/y). The moderate erosion means the rated of change between -1 to -5 (m/y). The high deposition and medium deposition mean the rates of change which are greater than +5, and +1 to +5 m/y, respectively. The stable coast shows low erosion or low deposition, where the rates of change are equal and less than 1 m/y (Table 2).

Status	Degree of coastal change	Average erosion rate (m/y)
Erosion	Severe erosion	> -5
	Moderate erosion	-5 to -1
	Low erosion (stable coast)	< -1
Deposition	High deposition	> +5
	Medium deposition	+1 to +5
	Low deposition (stable coast)	< +1

Table 2. The rate of coastal change

3. RESULTS AND DISCUSSION

3.1 Comparison of the coastline length

The length of the five coastlines including 2002, 2004, 2009, 2012, and 2016, was measured and compared. The coastline in 2002 showed the longest distance (8,361.48 m), followed by 2004 (7,792.38 m), 2012 (7,420.27 m), and 2016 (7,199.63 m). According to the data limitations of the images for the year 2009, which covered only the middle and lower parts of the study area, then the analysis of the year 2009 would cover based on available data. The length of the coastlines between the consecutive years was then compared and it was found that the changes in the coastline in the upper and middle parts were less severe than in the lower part of the island. The changes of the coastal length between 2002 and 2016 (1,161.85 m) showed the highest difference, higher than the coastline changes between 2002 and 2012 (941.21 m), 2002 and 2004 (569.10 m), and 2012 and 2016 (220.64 m). This indicated that there was a change in the coastline of the island during the study period.

3.2 The rate of coastal change

The rate of change in the study area showed that five locations of severe erosion were found between 2002 and 2004 (Figure 2). Between 2004-2009, it was found that eight locations of erosion were found, including one location with severe erosion and seven locations of moderate erosion. Ten locations of deposition were found, including two locations of high deposition, six locations of medium deposition, and two locations of low deposition (Figure 2). Between 2009 and 2012, it was found that five locations of erosion were found, including one location with severe erosion, three locations with moderate erosion, and one location with low erosion. Six locations of deposition were found, including two locations with high deposition, three locations with medium deposition, and one location with deposition (Figure 2). Between 2012 and 2016, seven locations of erosion were found, including one location with severe erosion, five locations with moderate erosion, and one location with low erosion. Eight locations of deposition were found, including seven locations with medium deposition and one with deposition (Figure 2).

3.3 The areas of coastal change between 2002 and 2016

The comparison of coastal changes between 2002 and 2004 showed that coastal erosion covering an area of 25.85 ha. Five locations were classified as severe erosion and covering an area of 25.76 ha. One location was classifies as low erosion and covering an area of 0.08 ha (Table 3 and Figure 2)

Year	Degree of erosion	Area (ha)
2002-2004	Severe	25.77
	Low	0.08
2004-2009	Severe	6.12
	Moderate	1.17
	Low	0.01
2009-2012	Severe	5.61
	Moderate	1.34
2012-2016	Severe	4.27
	Moderate	1.54
	Low	0.02

Table 3. Coastal erosion between 2002 and 2016

A comparison of the coastal changes during 2004 and 2009 showed that one location of severe coastal erosion was found, which covered an area of 6.12 ha. Seven locations with a moderate level of coastal erosion were found, which covered the area of 1.17 ha (Table 3 and Figure 2). In addition, two locations with low deposition covering an area of 2.91 ha were found, and six locations of medium deposition covering an area of 2.90 ha were found. Two locations with high deposition covering an area of 5.14 ha were found (Table 4 and Figure 2).

A comparison of the coastal changes between 2009 and 2012 showed that one location of severe erosion covering an area of 5.61 ha was found. Three locations of moderate erosion covering an area of 1.34 ha were found, and one location with low erosion covering an area of 0.01 ha was found (Table 3 and Figure 2). In addition, two locations with high deposition covering an area of 1.55 ha were found; and three locations with medium deposition covering an area of 1.20 ha were found. One location with low deposition covering an area of 0.02 ha was found (Table 4 and Figure 2).

Comparison of the coastal changes between 2012 and 2016 showed that one location with severe erosion covering an area of 4.27 ha was found, and five locations with moderate erosion covering an area of 1.54 ha were found. One location with low erosion covering an area of 0.02 ha was found (Table 3 and Figure 2). In addition, seven locations with medium deposition covering an area of 4.37 ha were found; and one location with low deposition covering an area of 0.04 ha was found (Table 4 and Figure 2).

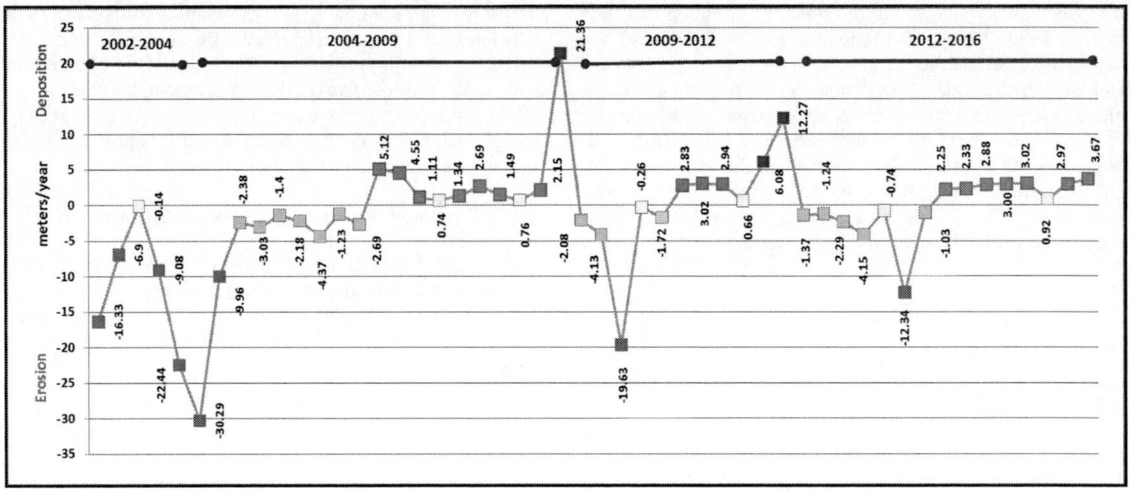

Figure 2. The rate of coastal change between 2002 and 2016

Year	Degree of deposition	Area (ha)
2004-2009	High	5.14
	Medium	2.91
	Low	0.03
2009-2012	High	1.55
	Medium	1.20
	Low	0.02
2012-2016	Medium	4.37
	Low	0.04

Table 4. Coastal deposition between 2002 and 2016

However, a comparison of the coastline changes between 2002 and 2016 for the 14-year period showed that one location with severe erosion covering an area of 28.14 ha was found. Three locations with moderate coastal erosion covering an area of 9.25 ha were found, and one location with low erosion covering an area of 0.07 ha was found. In addition, two locations with medium deposition covering an area of 6.27 ha were found, with one location with low deposition covering an area of 0.12 ha being found (Table 5 and Figure 2).

Degree of change	Erosion/ Deposition area (ha)	Rate of Change (m/y)
Severe	28.14	100.97
Moderate	8.21	36.48
Moderate	0.25	16.42
Moderate	0.79	15.55
Low erosion	0.07	4.98
Medium	6.21	26.37
Medium	0.05	14.71
Low deposition	0.12	13.9

Table 5. Coastal changes between 2002 and 2016

3.4 The risk areas in Ko Kho Khao

The erosion risk areas between 2002 and 2016 were evaluated and it was found that severe coastal erosion occurred in 2004. Then, natural adaptation demonstrated an important role in this area between 2004 and 2016, with the upper and middle parts of the area not showing the effect of severe erosion. In contrast,

the lower part of the island , especially the tip of the island , indicated severe erosion throughout the 14-year study period..

It indicated that there were other factors that caused the severe erosion. The study indicated that the coastal erosion was probably affected by the breakwater or seawall. It was found that the erosion areas between 2002 and 2004 exhibited a maximum erosion of 13.48 ha and higher than the erosion between 2004 and 2009 (6.12 ha), 2009 and 2012 (5.61 ha), and 2012 and 2019 (4.27 ha). The rate of erosion between 2002 and 2004 was 22.44 meters per year because of the tsunami phenomena. In addition, the rate of erosion between 2004 and 2009, 2009 and 2012, and 2012 and 2016 was 9.96, 19.63, and 12.34 meters per year, respectively (Table 6 and Figure 3).

It should be concluded that the first period of erosion from the tsunami disaster was in 2004. After that the erosion areas deceased because of natural recovery and adaptation. However, the tip at the lower part of the island was still found to have continuous severe erosion (Figure 3). The rate of erosion was 30.29 meters per year (2002-2004), 19.63 meters per year (2009-2012), and 12.34 meters per year (2012-2019).

Year	Year	Erosion area (ha)	Coastal length (m)	Rate of erosion (m/y)
2002-2016	14	28.14	2,787.01	100.97
2002-2004	2	13.48	3,004.36	22.44
2004-2009	5	6.12	1,228.88	9.96
2009-2012	3	5.61	952.21	19.63
2012-2016	4	4.27	865.58	12.34

Table 6. The coastal changes at lower part of island between 2002 and 2016

This research and previous research were compared, and the prediction of the study of Pantanahiran (2014) showed that the lower part of the island should have disappeared from erosion in 2015 (Figure 4), with the rate of erosion of 30 meters per year. However, the present study showed that some areas still remained as severe erosion was still active. This indicated that natural adaption plays a major role in this area; however, these areas are still in critical condition (Figure 5).

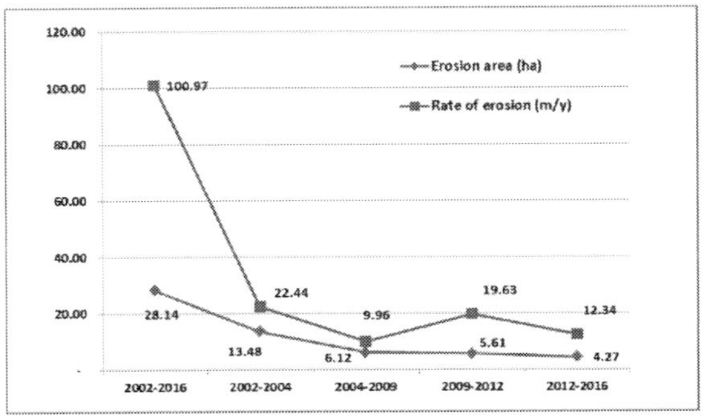

Figure 3. The coastal changes in the lower part of the island between 2002 and 2016

4. CONCLUSION

The results showed that the tsunami in 2004 caused coastal erosion in the area, as coastal changes during those years were found. However, natural adaptation was found after 14 years in the middle and upper parts of the island. Aerial photographs, IKONOS imagery, QuickBird imagery the Worldview-2 imagery, and Pleiades imagery were used. These remote sensing data were very useful tool for the coastal change. Severe coastal erosion in the lower part of the island has been continuously found, and it was also found that the seawall might be the main factor in the coastal erosion after tsunami occurrence. It is recommended that the lower part of the island be declared a special observation area in order to prevent further coastal erosion.

ACKNOWLEDGEMENTS

This work was supported by the Higher Education Research Promotion and National Research University Project of Thailand, Office of the Higher Education Commission. The author would like to extend his appreciation to the Strategic Wisdom and Research Institute, and the Faculty of Social Science at Srinakharinwirot University, for their support and encouragement.

REFERENCES

Clark, R., 1996. *Coastal zone management*. CRC Press, Inc. Boca Raton, Florida, USA. pp. 236.

Department of Mineral Resources, 2006. *Survey study to restore and develop coastal areas Ban Nam Khem Takua Pa Phang Nga province*. Ministry of Natural Resources and Environment. Bangkok, Thailand (in Thai).

Kennedy, H., 2001. *Dictionary of GIS Terminology*. The ESRI Press. Redlands, CA, USA.

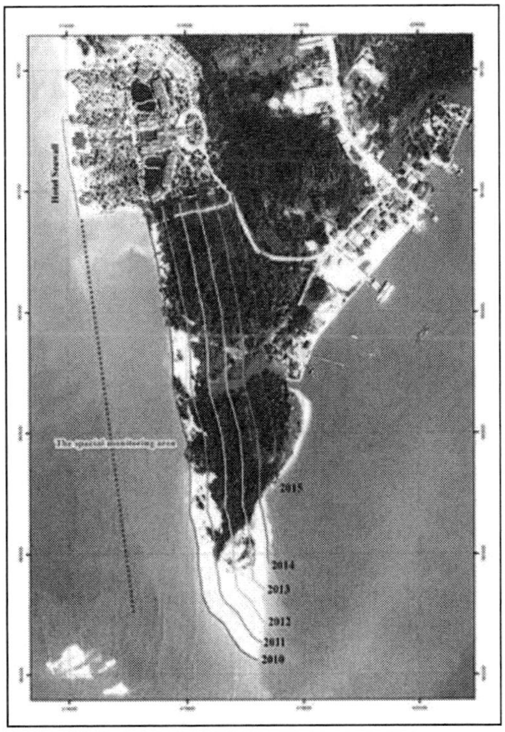

Figure 4. The prediction of previous study of Koh Kho Khao (Pantanahiran 2014)

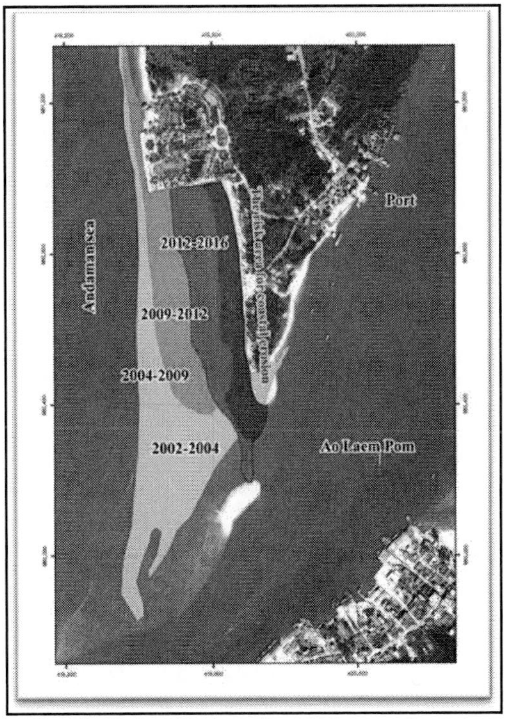

Figure 5. The severe erosion of Koh Kho Khao and the special monitoring area

https://doi.org/10.5194/isprs-archives-XLII-3-W7-51-2019

ONEP, 2003. *Mater plan study for coastal erosion alleviation from Pethchaburi river mouth to Pranburi river mouth*. Bangkok, Thailand (in Thai).

Pantanahiran W., Weesakul S. and Thaicharoen C., 2008. Monitoring the coastal change in Thailand using GIS technology. In: *Proceeding of the 7th International Conference of Asia GIS 2008*. Pusan, Republic of Korea.

Pantanahiran, W., 2014. *Monitoring the Coastal Change after Tsunami in Thailand.* IOP Conf. Ser.: Earth Environ. 17 012108, doi:10.1088/1755-1315/17/1/012108.

Sinsakul S., et al., 2003. *The technical report: The coastal change of Andaman sea*. Department of Geology, Bangkok Thailand (in Thai).

Mapping the spatial distribution of Colombia's forest aboveground biomass using SAR and optical data

P. Rodríguez-Veiga [1, 2,*], A.P. Barbosa-Herrera[3], J.S. Barreto-Silva[3], P. C. Bispo[1], E. Cabrera[3], C. Capachero[3], G. Galindo[3], Y. Gou[1], L.M. Moreno[3], V. Louis[1], P. Lozano [3], A. M. Pacheco-Pascagaza [1], I. P. Pachon-Cendales[3], J.F. Phillips-Bernal[3], J. Roberts[1], N. R. Salinas[3], L. Vergara[3], A.C. Zuluaga[3], H. Balzter[1,2]

[1] Centre for Landscape and Climate Research, University of Leicester, Leicester, UK - prv4@leicester.ac.uk
[2] NERC National Centre for Earth Observation (NCEO), Leicester, UK
[3] Institute of Hydrology, Meteorology and Environmental Studies (IDEAM), Bogota, Colombia - pbarbosa@ideam.gov.co

Commission III

KEY WORDS: biomass, carbon, SAR, multispectral, machine-learning, forest inventory

ABSTRACT:

An assessment on the amount and spatial distribution of forest aboveground biomass (AGB) for the forests in Colombia was generated using in-situ national forest inventory data (IDEAM, 2018), in combination with multispectral optical data and synthetic aperture radar (SAR) satellite imagery. ALOS-2 PALSAR-2 gamma-0 backscatter annual mosaics (2015-2017) provided by JAXA were normalised and corrected using previous ALOS PALSAR annual mosaics (2007-2010) as reference. A multi-temporal Landsat 7 & 8 composite over the whole of Colombia was used for the year 2016 ± 1. The national forest inventory in-situ plots used to train our model consisted of 5-subplots each and were collected during the period 2015-2017 in the main biomes of the country. A sample of permanent 1ha plots (PPMs) were also measured. Nationally developed allometries (Alvarez et al, 2012) were used to estimate AGB. A non-parametric random forests (RF) algorithm was used within a k-fold framework to retrieve AGB at 30m spatial resolution for the whole of Colombia. The algorithm was trained using forest inventory plots and validated at plot (0.35 ha) and PPM level (1 ha). The accuracy assessment found coefficients of determination (R^2) of 0.68 and 0.61, and relative root mean square errors (Rel. RMSE) of 49% and 34% at plot and at PPM level, respectively. The results showed that the average AGB for the country was 118.1 t ha^{-1} (45.6 t ha^{-1} for Caribe, 75.4 t ha^{-1} Andes, 122.5 t ha^{-1} Pacifico, 32.7 t ha^{-1} Orinoquia, and 200.5 t ha^{-1} for the Amazonia, regionally), and that the total carbon stocks for the country were 6.7 Pg C for the period 2015-2017.

1. INTRODUCTION

Forest cover area in Colombia amounts to 52% of the national territory. Forests are one of the most important resources in terms of goods and ecosystem services, which are vital for the sustainable development of the country. Thus, forest monitoring is essential for decision-making processes related to forest management and reduction of deforestation.

The government of Colombia signed the Joint Declaration of Intent in 2015 with the governments of Norway, Germany and the United Kingdom to cooperate in reducing emissions from deforestation and forest degradation (i.e. REDD+) and promoting sustainable development in Colombia. In the last years, Colombia has strengthened its monitoring capabilities by developing the National Forest Information System, the Forest and Carbon Monitoring System, and the National Forest Inventory (NFI). Those have allowed to improve the monitoring of forest dynamics associated to disturbances leading to greenhouse gas emissions, and to promote the conservation and sustainable management of forests.

Aboveground biomass (AGB) is an Essential Climate Variable (ECV) required by the International Panel on Climate Change (IPCC) to carry out greenhouse gas inventories of forest ecosystems. AGB is the basic unit to account for carbon stored in vegetation. Activities aiming at the reduction of greenhouse gas emissions from deforestation and forest degradation require information on the spatial distribution of the carbon stocks. This information can be used to evaluate the carbon gains due to new forest or growth, as well as the loss due to forest disturbances such as deforestation or degradation.

The National Forest Inventory of Colombia started in 2015 and it is an initiative aiming to monitor forest resources nationwide (IDEAM, 2018). The NFI collects information on vegetation structure, composition, richness, biodiversity, biomass, growing stock, soil carbon, and forest dynamics. This information allows to monitor the status of forests, the carbon stocks, and the composition of forest resources through time. However, the number of sampling units to date is quite limited due to difficulty to access the most remote forest areas (i.e. Amazon region).

The aim of this study is to estimate the amount and spatial distribution of Colombia's AGB using the current NFI sample and Earth Observation (EO) data. Then, the results will be compared with the current NFI estimates at national and regional level to understand where the main discrepancies are.

* Corresponding author

https://doi.org/10.5194/isprs-archives-XLII-3-W7-57-2019

2. MATERIALS

2.1 National Forest Inventory (NFI)

The NFI uses two types of sampling units: plots and 1-ha square plots (PPMs). Approximately 300 plots and 26 PPMs have been measured since 2015 (IDEAM, 2018). However, the Amazon region still presents very sparse sampling in the core areas (Figure 1).

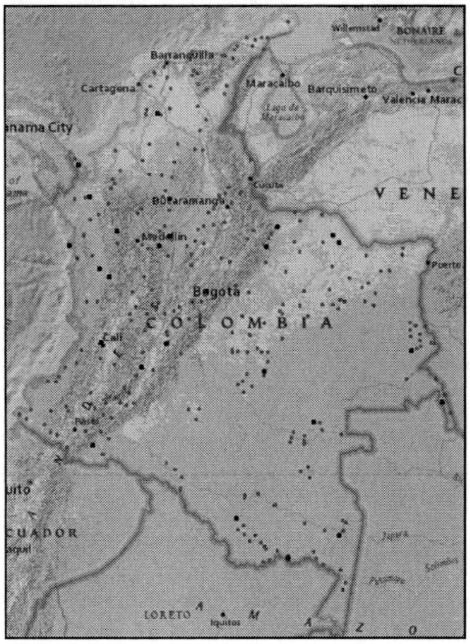

Figure 1. Location of plots (red dots) and PPMs (black squares)

A NFI plot is composed of five 30 m diameter circular subplots (SPFs) (Figure 2). Thus, each plot consists of 0.35 ha of actual measurements (0.07 ha per SPF). The dimensions of the PPMs are 100m x 100m and usually share the central coordinates with the centre of the central SPF (SPF-1) within an existing plot (Figure 2). Additionally, the SPF follow a concentric nested sampling for different diameter classes (Figure 2).

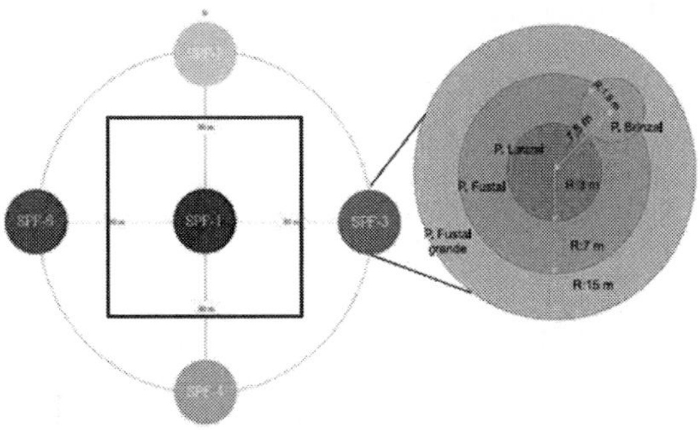

Figure 2. Structure of plots and 1-ha square plots

The NFI plots are used as the main training/validation sampling unit in this study, while PPMs provided an additional validation scenario. The size and design of the PPMs would be more optimal than the plots for using with EO data (Réjou-Méchain et al., 2014), but the reduced number of those limits its current use. A 10 m buffer was applied in this study around each 30 m SPF in order to reduce location errors. This resulted in actual 50 m diameter subplots (~0.2 ha per subplot x 5 = 1 ha per plot) used to sample the overlapping pixels from the EO datasets for the training of the algorithm.

The NFI calculates carbon stocks nationally and regionally using the ratio estimator frequently used for the Global Forest Resources Assessment reporting (Marklund et al., 2005).

2.2 Earth Observation Datasets

A combination of multispectral and SAR datasets was used in this study. The JAXA's ALOS-2 PALSAR-2 freely available 25m resolution mosaics with standard SAR pre-processing (Shimada et al., 2014) were used. This processing consists of calibration, multi-looking (output of 16 looks), projection, ortho-rectification, slope correction using SRTM DEM and an additional destriping process (Shimada and Isoguchi, 2002). We additionally process these mosaics with the following steps: 1) removal of problematic pixels based on quality layers, 2) multi-temporal filtering (5x5 window), 3) normalization and correction based on temporal statistics of ALOS PALSAR previous mosaics, and 4) gap-filling. Landsat 7 and 8 multispectral 30m spatial resolution normalized top of atmosphere (TOA) annual composites (Hansen et al., 2013) were also used. The multispectral dataset Proba V 100m (Dierckx et al., 2014), and the 30m Digital Elevation Model from the Shuttle Radar Topography Mission (SRTM) were also explored during the variable selection stage.

3. METHODS

3.1 Allometric models

Non-destructive sampling of tree level variables such as diameter at breast height (*D*), canopy height (*H*), and species identification were collected from trees, palms and shrubs. AGB was calculated for trees using the pantropical allometric models of Chave et al. (2014) and the national models of Alvarez et al. (2012), and then compared. In both cases the allometric models with *D* as sole predictor variable were used. In the case of palms and shrubs, the Tiepolo et al. (2002) allometric models were used.

3.2 Algorithm

The implementation of the non-parametric machine learning algorithm Random Forest (Breiman, 2001) in the Google Earth Engine platform was used. The regression version of the algorithm was implemented within a k-fold framework to maximize the number of plots available for training and validation of the algorithm (i.e. 300 plots). Therefore, all the data was used for training and validation purposes. The dataset was stratified by three AGB levels, and randomly sampled into the folds to ensure that all folds have similar probability distribution functions of AGB.

The EO input variables for the algorithm were selected based on quantitative analyses (i.e. jackknife analyses) as well as qualitative observations (minimum presence of artefacts in the resulting map).

https://doi.org/10.5194/isprs-archives-XLII-3-W7-57-2019 | © Authors 2019. CC BY 4.0 License.

4. RESULTS

The allometric models from Chave et al. (2014) derived in a 11% higher AGB estimate in comparison to Alvarez et al. (2012) models (average AGB of 125 t ha^{-1} vs. 113 t ha^{-1}) (Figure 3). The national models from Alvarez et al. (2012) were finally used to develop the AGB map.

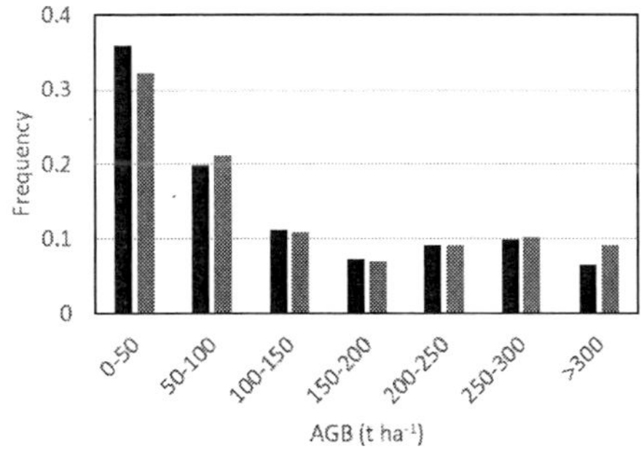

Figure 3. Histograms of AGB estimates from NFI plots using Chave et al. (2014) and Alvarez et al. (2012) models

The results of the Jackknife analyses based on root mean square error (RMSE), bias, and coefficient of determination (R^2) indicated that the combination of ALOS-2 PALSAR-2 and Landsat variables showed the best balance between the lowest possible bias and RMSE, and the highest possible R^2 (Figure 4)

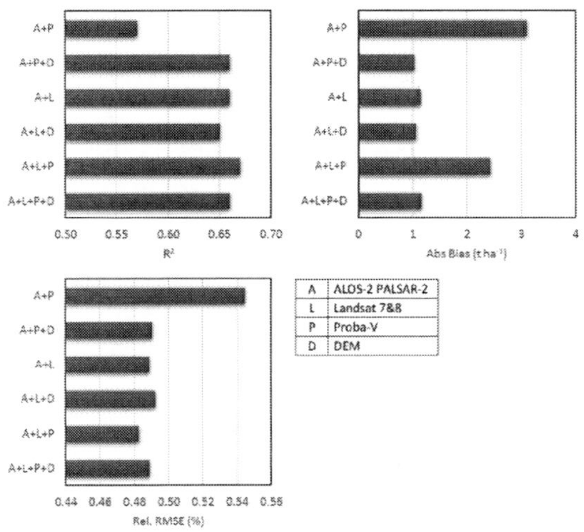

Figure 4. Jackknife analyses based on R^2, bias and RMSE for the potential input variables

The AGB map was therefore generated using ALOS-2 PALSAR-2 and Landsat 7 & 8 composite variables at 30 m spatial resolution for the whole of Colombia (Figure 5).

The average AGB density estimated for the country using the AGB map was 118.1 t ha^{-1}. By regions, the average AGB density values were 45.6 t ha^{-1} for Caribe, 75.4 t ha^{-1} for Andes, 122.5 t ha^{-1} for Pacifico, 32.7 t ha^{-1} for Orinoquia, and 200.5 t ha^{-1} for the Amazonia. The total forest carbon stock for the country was estimated in 6.7 Pg C for the period 2015-2017. The Amazon region has more than half of the AGB stocks in the country.

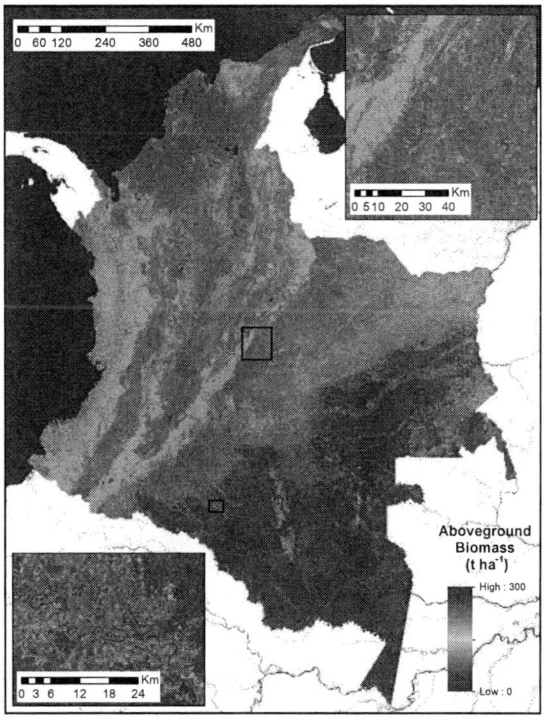

Figure 5. Colombia AGB map for year 2016±1

The R^2 of the final map was 0.68, the Rel. RMSE 49%, and the bias -2.6 t ha^{-1} (Figure 6). The additional validation using the PPMs showed similar results with R^2=0.61, Rel. RMSE=34% and Bias=-7.1 t ha^{-1}.

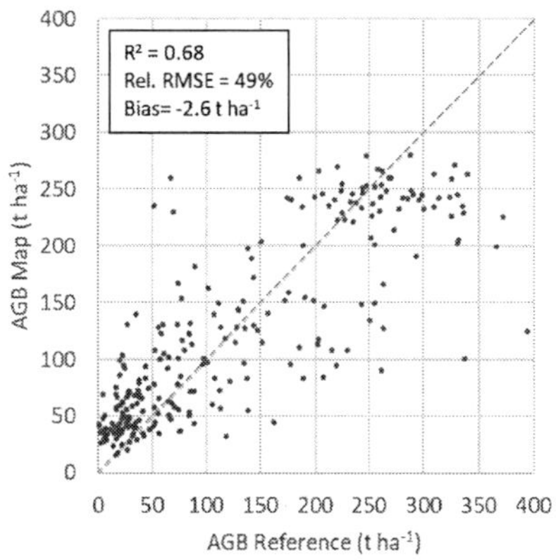

Figure 6. Scatterplot AGB map vs. AGB reference

The AGB stocks calculated using the AGB map were comparable to current estimates from the NFI. Table 1 shows that although the AGB estimates derived from the map are lower than the NFI estimates, the differences are not significative. The total AGB estimate for the whole country was only a 2.93% lower than the NFI national estimate, while regionally the differences were between 0.87% and 28.19% (Table 1).

Table 1. Comparison between AGB map estimates and NFI estimates by region

Region	Total AGB (Pg)	Difference to NFI (Pg)	Difference to NFI (%)
Amazonía	9.19	1.453	13.65
Andes	2.20	0.252	10.28
Caribe	0.70	0.006	0.87
Orinoquía	0.55	0.216	28.19
Pacífico	0.82	0.147	15.14
Total	**13.47**	**0.406**	**2.93**

5. DISCUSSIONS

The AGB plot data calculated using Alvarez et al. (2012) allometric models showed 11% lower values than AGB data calculated by Chave et al. (2014) models. These differences could be related to the different sampling datasets used to develop the models per se (national vs. pantropical). However, it can also be related to the use of the simple models using *D* as predictor variable instead of the models using *D, H*, and specific wood gravity (ρ). The AGB map was finally derived by training the algorithm with AGB plot data derived from Alvarez et al. (2012) allometric models.

The largest difference between the AGB map estimates and the NFI estimates occurs in the Orinoquia and Pacifico regions. Those are the regions with the lowest number of plots available from the NFI and the differences might indicate the need for more sample plots. Additionally, even though Orinoquia region is mostly covered by savannah vegetation (forest represents only 13% of the area), more than half of the plots sampled in that region were located in forests. This might lead to an overestimation by the NFI in that region using the ratio estimator method and explain the large difference found with the AGB map in that region (28.19%). Caribe was the region with the highest agreement to the AGB map estimates with just a 0.87% difference. This region had the highest sampling per unit of area, which

6. CONCLUSION

This is the most detailed and accurate map to date produced for Colombia using national data. Initial results using a combination of SAR ALOS-2 PALSAR-2 and multispectral Landsat 7 & 8 annual composites show promising results. Future work however should explore the use of allometric models with *D, H*, and ρ as predictors to generate more accurate AGB plot data to train the algorithm, as well as larger plots more suited to use in combination with EO data.

ACKNOWLEDGEMENTS

The authors would like to thank JAXA, NASA, and USGS for freely providing the EO data. Thanks also to Google for providing the processing platform (Google Earth Engine). This research was supported by the Forests 2020 project. The Forests 2020 project is funded by the UK Space Agency International Partnership Program. P. Rodríguez Veiga and H. Balzter were supported by the UK's National Centre for Earth Observation (NCEO).

REFERENCES

Alvarez E, Duque A, Saldarriaga J, Cabrera K, de las Salas G, del Valle I, et al. Tree above-ground biomass allometries for carbon stocks estimation in the natural forests of Colombia. Forest Ecology and Management. 2012;267(0):297-308.

Breiman, L. 2001. Random forests. Machine Learning, 45, 5-32.

Chave, J., Réjou- Méchain, M., Búrquez, A., Chidumayo, E., Colgan, M. S., Delitti, W. B., Duque, A., Eid, T., Fearnside, P. M. & Goodman, R. C. 2014. Improved allometric models to estimate the aboveground biomass of tropical trees. Global change biology, 20, 3177-3190.

Dierckx, W., Sterckx, S., Benhadj, I., Livens, S., Duhoux, G., Van Achteren, T., Francois, M., Mellab, K. & Saint, G. 2014. PROBA-V mission for global vegetation monitoring: standard products and image quality. International journal of remote sensing, 35, 2589-2614.

Hansen, M. C., Potapov, P. V., Moore, R., Hancher, M., Turubanova, S. A., Tyukavina, A., Thau, D., Stehman, S. V., Goetz, S. J., Loveland, T. R., Kommareddy, A., Egorov, A., Chini, L., Justice, C. O. & Townshend, J. R. G. 2013. High-Resolution Global Maps of 21st-Century Forest Cover Change. Science, 342, 850-853.

IDEAM (Institute of Hydrology, Meteorology and Environmental Studies). National Forest Inventory. Url: http://www.ideam.gov.co/web/ecosistemas/inventario-forestal-nacional. 2018.

Marklund, L.G. and Schoene, D.I.E.T.E.R., 2006. Global assessment of growing stock, biomass and carbon stock. Forest Resources Assessment Programme Working paper, 106.

Réjou-Méchain, M., Muller-Landau, H. C., Detto, M., Thomas, S. C., Toan, T. & Saatchi, S. S. 2014. Local spatial structure of forest biomass and its consequences for remote sensing of carbon stocks. Biogeosciences, 11.

Shimada, M. & Isoguchi, O. 2002. JERS-1 SAR mosaics of Southeast Asia using calibrated path images. International Journal of Remote Sensing, 23, 1507-1526.

Shimada, M., Itoh, T., Motooka, T., Watanabe, M., Shiraishi, T., Thapa, R. & Lucas, R. 2014. New global forest/non-forest maps from ALOS PALSAR data (2007–2010). Remote Sensing of Environment, 155, 13-31.

Tiepolo, G., Calmon, M., & Feretti, A. R. 2002. Measuring and monitoring carbon stocks at the Guaraqueçaba climate action project, Paraná, Brazil. In Proceedings of the International Symposium on Forest Carbon Sequestration and Monitoring (pp. 11-15). Taipei: Taiwan Forestry Research Institute.

Revised January 2019

STUDY ON EXTRACTION METHODOLOGY OF RICE GROWTH CHARACTERISTICS IN FIELD USING TERRESTRIAL LASER MEASUREMENT DATA

K. Sakata [1], A. Rikimaru [1], Y. Yasunaga [2],

[1] Dept. Civil and Environmental Engineering, Nagaoka University of Technology, 1603-1 Kamitomioka, Nagaoka, Niigata, Japan - sakata@vos.nagaokaut.ac.jp

[2] Ecology Science co., ltd, 3-5-25 Kusoudzu Nagaoka, Niigata, Japan - yasunaga@zb.wakwak.com

Commission III, WG III/10

KEY WORDS: Terrestrial laser scanner, rice plant height, point clouds,DSM

ABSTRACT:

Smart agriculture is recommended for labor-saving, refinement and high quality production of rice cultivation with safety and security. The Terrestrial laser measurement has the advantage of acquiring 3D point cloud information with high accuracy and high density, but there is a disadvantage that it becomes difficult to measure the base of rice plant when paddy rice grew. In this study, the 3D point cloud information obtained by terrestrial laser measurement was used to three-dimensionally measure the growth state in the field corresponding to the growing period in time series. The growth characteristics of the whole field as a community were extracted as a representative value of mean growth. Then we attempted extraction of individual differences in growth and growth unevenness, as the deviation from the mean value of the field in terms of the growth characteristics of each strain focused on the individual plants. it is difficult to observe the root of the rice during the growing period, it was used for analyzing the growth quantity as the growth difference height for each growing period by obtaining the difference of the elevation value of the rice at the early stage of growth and at each stage of growing. The spatial variation in the growth quantity of the whole field area and the local area was calculated, the range of the deviation for evaluating the mean variation and the growth abnormality was examined, and the setting of the threshold for evaluating the growth normal and abnormality was examined.

1. INTRODUCTION

In Japan, smart agriculture aiming at maximum production yield, production control using ICT technology advanced was recommended. Remote sensing technology has attracted attention for precise cultivation related to smart agriculture. For example, Inoue and Yokoyama (2017) has proposed a field crop and farmland diagnosis using a drone equipped with an optical sensor. N. Tilly et al. (2014) suggests estimating biomass and height of corn using a terrestrial laser scanner. In this research, we are aiming to grasp the growth condition of paddy rice using Terrestrial Laser Scanner (TLS) which is excellent in portability and can acquire three dimensional information in a wide range. It suggested that the growth survey is possible using the surface height of paddy rice by TLS. The bottom end and the top end of paddy rice detected from the TLS viewed directly under and evaluated as rice plant height, indicating that its effectiveness is high. In this research, using this TLS, we examined the extraction method of rice growth characteristics considering the influence of measurement characteristics from oblique measurement. As growth characteristics, paddy rice has individual differences in growth, and it is useful to detect the difference in growth in its solid difference. Therefore, by using TLS that can be precisely measured, it is possible to extract growth characteristics using values related to plant height.

In presently, acquisition method of rice growth information was mean value that measured ten plant of plant height, leaf color, stem number in a field area.

Currently, as for the method of acquiring the growth information, 10 plants height, leaf colour, stem number are measured for one field, and they are averaged to represent the field value.

Examination of the total number of fields is difficult due to human factors, but it is useful as a simple representative value extraction. However, if there are individual differences even in the 10 stocks to investigating, the representative values of the extracted rice stocks are dangerous in the whole field.

Therefore, we examined a method of extracting information on the growth characteristics of rice according to the index value without ground effect using TLS is excellent in portability and can measure area. In TLS measurement, there is no human measurement error, it has objectivity and reproducibility, and it is possible to conduct a survey of all stocks through a planar observation. Based on this planar measurement information, we attempted a method of extracting growth information taking individual differences of paddy rice into consideration.

2. METHODS

2.1 Study area

The study site is located at a paddy field of Nagaoka Niigata, Japan (N37°25'3'', E138°47'50'').

In this study, 3D point clouds were acquired using RIEGL VZ - 400. In order to obtain the point group in detail, the observation pitch angle was set to 0.02 [°]. The TLS measurement is from the rice planting stage to the harvesting stage and is carried out on June 21, July 1, July 11, July 19, July 29, August 9, August 23, September 13 2016, it measured eight times in total. Manual measurement to measure plant height, plant stem, and leaf color on the same day.

In order to extract the growth characteristics, growth conditions were managed by fertilizer and grown. The classification of fertilizing quantity is shown in Fig.1. The A compartment applied normal fertilizer to the basic fertilizer, the B compartment set the

This contribution has been peer-reviewed.
https://doi.org/10.5194/isprs-archives-XLII-3-W7-61-2019 | © Authors 2019. CC BY 4.0 License.

basic fertilizer to half of usual, and there was no basic fertilizer in the C compartment. We applied fertilizer as usual in A1, B1, C1 compartment, half of fertilizer was applied in A2, B2, and fertilizer was not applied in C2 compartment. The A1 compartment is a standard fertilizer application. Rice plants are May 16th and additional fertilizer is July 22. Paddy field measurement by TLS was set 6 points, and 6 points data were composited. The standard deviation of the height of each observation day was about 2 mm with the horizontal board installed at the four corners of the field.

Fertilization	Base Fertilizer [kg]	Additional Fertilizer [kg]
A1	3.00	1.92
B1	1.50	1.92
C1	0.00	1.00
A2	3.00	1.00
B2	1.50	1.00
C2	0.00	0.00

The Quantity of Nitrogen per 1000m^2

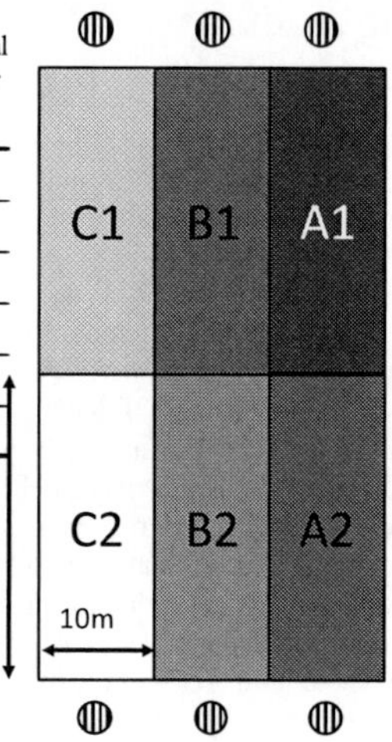

TLS measurement points

Fig.1 Experimental site and fertilization blocks

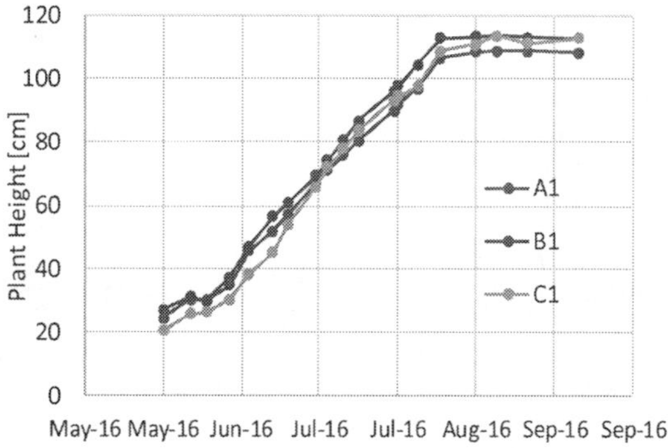

Fig.2 manually measured rice plant height

2.1 Generating DSMs

TLS measurements were taken at six points from the field fairway, and the six data were combined.
For the conversion to the DSM raster image, the rice point group data acquired by TLS was taken as the 99% tile value in the 5 cm mesh. In order to remove noise components above the paddy field, the tile value was 99%.

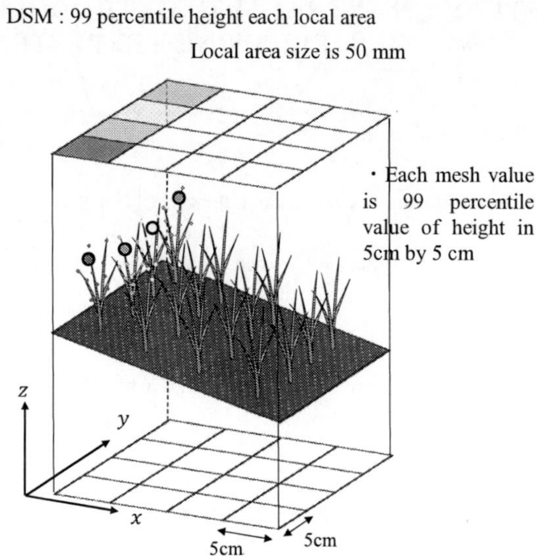

Fig.3 Experimental site and fertilization blocks

3. RESULT

Extraction method of rice growth characteristics using 3D point cloud from TLS measurement is below.

- Calculating rice different plant height
- Extraction rice growth characteristics

3.1 Calculation of rice different plant height

In the oblique measurement from TLS used in this research. The laser is obliquely irradiated to paddy rice, when it grows, due to the shielding of the ground by the leaf surface, the influence of the inclination and roughness of the ground, the practice with measurement; it is difficult to extract the height from the ground from the three-dimensional point cloud information of TLS.
Phan et al (2016) have constructed a regression model with plant height using the relative vertical distance of rice with laser measurement data in direct viewing. In the calculation of the relative vertical distance, when the shielding effect by the leaf surface is high at time of early growth period, the lower layer became unstable. Therefore, as shown in the Fig 4, the height difference value was calculated with respect to the change in the surface layer height of the rice bowl with the measurement height at the initial stage of growth as the reference height. By using this method, it is also possible to calculate the growth quantity without considering the inclination and irregularities of the ground. In this study, we analyzed the rice differential difference RDH as the growth quantity for each observation day. The formula for calculating RDH is shown in Equation 1.

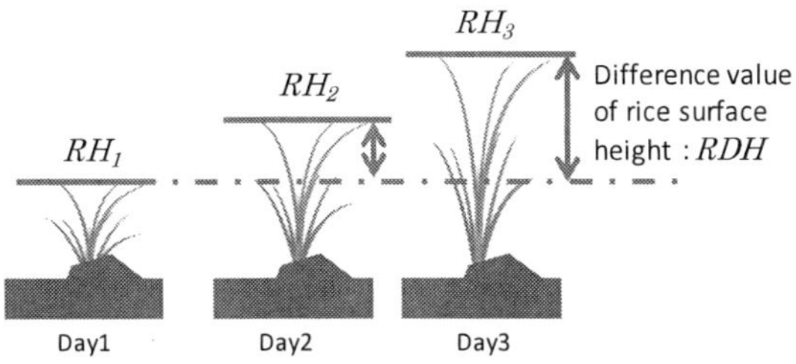

Fig.4 Image of RDH

$$RDH = RH_i - RH_1 \qquad (1)$$

where RDH = Rice Different Height [m]
RH = Rice Height on local area [m]
i = measurement day

The image of RDH calculated based on the first observation date is shown in Fig 5.
The image on September 13 has areas where rice is lodged and RDH is calculated low.

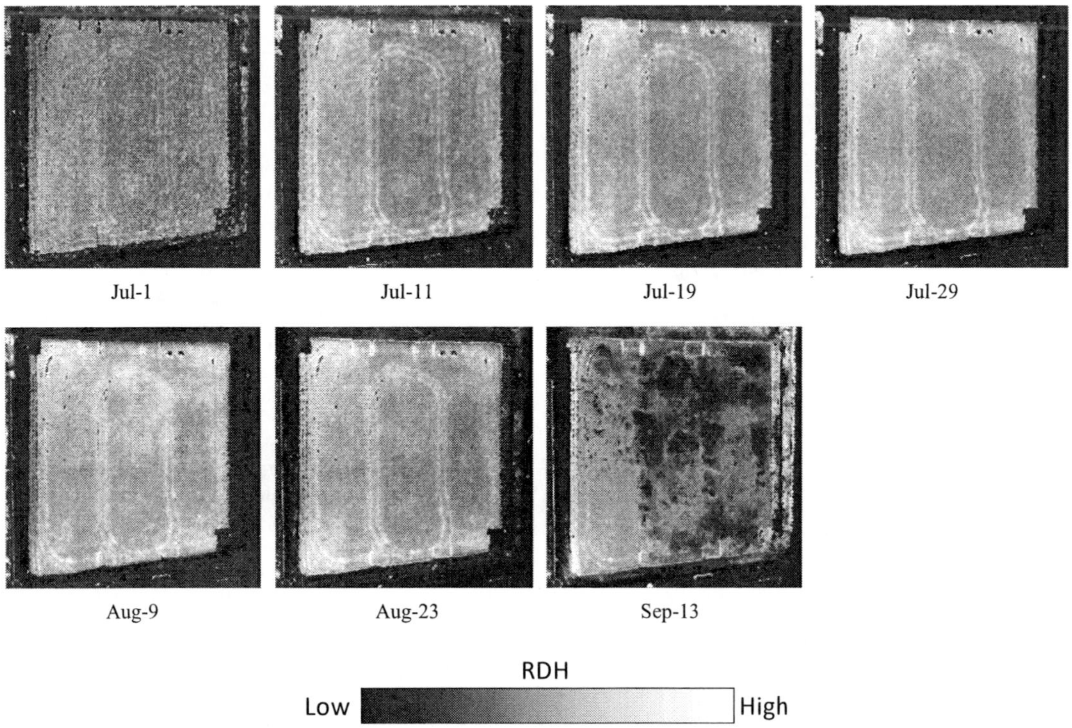

Fig.5 Images of RDH

For verification of RDH, it was checked against PDH calculated from practice measurement. For the calculation of the difference value, it is necessary to detect the local highest surface layer height RH of the rice plant. However, when growing, the position of RH changes within the range of one share depending on the curvature of the foliage and the way in which the rice plant collapses. Therefore, in order to cope with the change in the travel distance of RH, we developed a matching method in this research. Fig.6 shows the method of searching for the highest surface layer in the matching method.
As shown in Fig. 6, the initial place of practice measurement was set, and the highest value within the search window centered on that place was extracted. The maximum value extracted was calculated as the surface layer height at the observation day as the difference value of the surface layer height from the beginning. In this study, it is assumed that the search window falls within the search window that is compatible with the inter-strain and inter-section, and the maximum height is the plant height with one stock. In the target field, the distance between plants is about 20 cm and the distance between strands is about 30 cm, so the

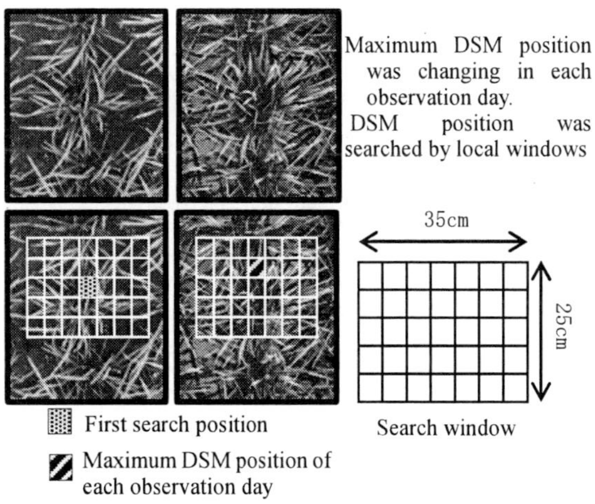

Fig. 6 Plant height search method

search range is set to 25 cm × 35 cm. For practical measurement, the value of plant height was used to obtain the plant height difference value from the beginning. The differences between the plant height difference value and the manual measurement plant height were checked each other. The manual measurement plant different height was calculated by Equation 2. Fig.7 shows the RDH and PDH excluding September 13, which is calculated by this method and is the ripening period. The reason for excluding the ripening stage is to measure the height of the crown when the rice plant is lodged when it is ripened, whereas in practice measurement the rice cultivation stands and the difference is shown.

$$PDH = PH_i - PH_1 \quad (2)$$

where $PDH = Plant\ height\ different\ height\ \ [m]$
$PH = manual\ measurement\ plant\ height\ [m]$
$i = measurement\ day$

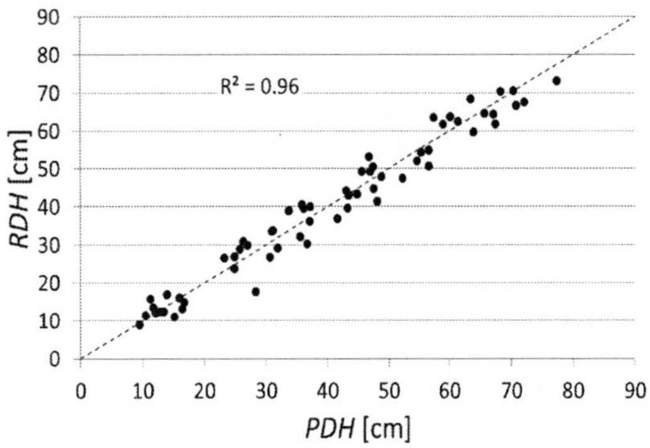

Fig.7 Compare RDH and PDH

The coefficient of determination between RDH and PDH was 0.96. The RMSE of PDH and RDH was 3.5 cm.

3.2 Extraction rice growth characteristics

The variation in growth of each fertilizing area is defined as Rice growth unevenness. In this paper, fertilization conditions are different for each area, and due to the difference in growing situation, we compare the variation of the growing state.
The growth situation is judged by measuring the individual stock. However, depending on the measurement place of the individual stock, there is a danger to the representative value of the field because there are individual differences. Therefore, by using TLS, we examined the extraction method of the growth characteristics using the variation about the difference of the growth. As shown in the Fig.8, when rice plants are delayed in growth due to individual differences and there are small rice stocks or rice stocks growing quickly and growing large, on the contrary from the average value of the standard growing plant height, the plant height becomes small or large. We propose a method to judge the growth condition of the plant height by using the standard deviation showing the variation. The histogram for each fertilizing area is shown in Fig.9. It understands that the value of RDH increases with each growth in each section. The threshold value is set by using μ and σ. Differences in growth are judged from the histogram of fertilization conditions of each section and used as a threshold.
It was assumed that difference in growth due to individual differences appear spatially randomly, and growth unevenness

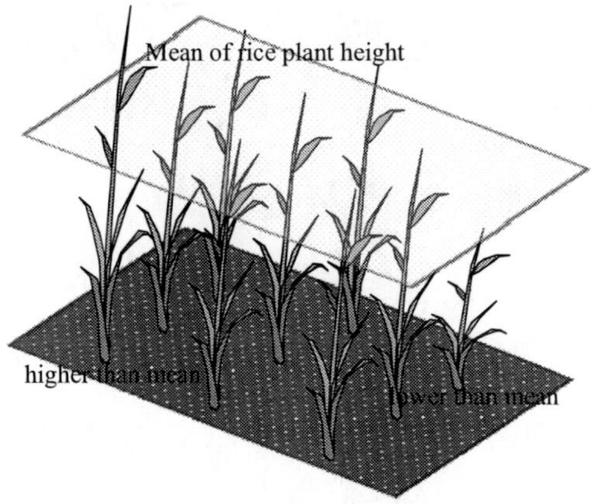

Fig.8 Image of rice growth unevenness in a rice field

due to soil condition and fertilization conditions appears spatial clusters.

3.3 Distribution images of growth characteristics

Distribution images of growth characteristics was prepared using average value μ and standard deviation σ. Fig.10 shows the variation of growth using standard deviation. The average growth is gray color.
From Fig.10, small plant height regions in fertilization area are distributed in spatially cluster, so growth unevenness could be extracted. On July 1, the low growth area and the high growth area are sparsely distributed, but after July 11, the high area and the low area are solidified and distributed. It suggests that the growth difference gradually appears after July 11.
The histogram of the fertilization area is shown in Fig.11. Regarding threshold setting Fig.11 A1 shows the histogram of A1 which is a standard habitat environment and C2 which does not give any fertilizer. The mode by RDH was almost the same in A1 and C1 section. In addition, separation was difficult even with the value of μ ± σ.

4. DISCUSSION

In this paper, we investigated the feature extraction method of growing condition due to the planar variation of field using TLS. In shown histogram of each fertilization area, since there was not much difference in standard deviation in each fertilization area, there was no change in growth unevenness. Fig.12 shows time series of RDH. In the C1 and C2 sections without the base fertilizer, the slope of the community RDH during the busy period is large with respect to the A section and the B section. However, from the practice measurement, the height of the plant is larger in the A section and B section. In this study, the RDH is calculated as the difference value from the initial plant height, but if the growth until this initial observation day is large, the RDH after that tends to decrease. In this study, it was caused by the fact that it was growing at the early observation days in the A and B plots. On the other hand, because there was no base fertilizer in the C1 and C2 area, there was a delay in growth at the initial stage, and RDH was calculated largely because it grew after the initial measurement. We attempted to extract growth characteristics using μ ± σ, but from the histogram of Fig. 12, it became difficult to detect a clear threshold value due to the influence of initial observation. However, it was suggested that by using μ ± σ, it is possible to grasp the distribution of growth variation in each section. In addition, according to the distribution map, it is possible to understand areas where growth delay and growth are earlier than average growth. It is understood that the region where growth is gradually increasing is gradually increasing.

5. CONCLUSION

We investigated a method of extracting variation of growth in a field with different fertilization conditions by using RDH index without influence of ground height. It was suggested that by using $\mu \pm \sigma$, it is possible to grasp distribution such as growth variation in each section. Growth unevenness was extracted in spatially cluster area. For the RDH, it is found that the timing of the initial observation becomes important. In the histogram of A1 and C2, too much difference was not confirmed in the variation. Growth diagnosis with only plant height is dangerous, and comprehensive judgment is required such as leaf color and number of stems.
It was confirmed that the variance was small in the whole section by conducting measurement survey of all plant height using laser.

REFERENCE

Yoshio Inoue, Masaki Yokoyama, Drone-based Remote Sensing of Crops and Soils and its Application to Smart Agriculture, Journal of the Remote Sensing Society of Japan, Vol.37, No.3,224-235(2017)

N. Tilly et al, Evaluation of terrestrial laser scanning for rice growth monitoring , Inc. Arch. Photogramm. Remote Sens. Spat . Inf. Sci. XXXIX-B7 , 351-356 (2012)

N. Tilly et al ,Terrestrial Laser Scanning For Plant Height Mesurement and Biomass Estimation of Maize , Inc. Arch. Photogramm. Remote Sens. Spat . Inf. Sci. VII-7 , 181-187 (2014)

Anh Thu Thi Phan et al, Method for estimating rice plant height without ground surface detection using laser scanner measurement, J.Appl.Remote Sens., 10(4), 046018(2016), doi:10.117/1.JRS.10.046018

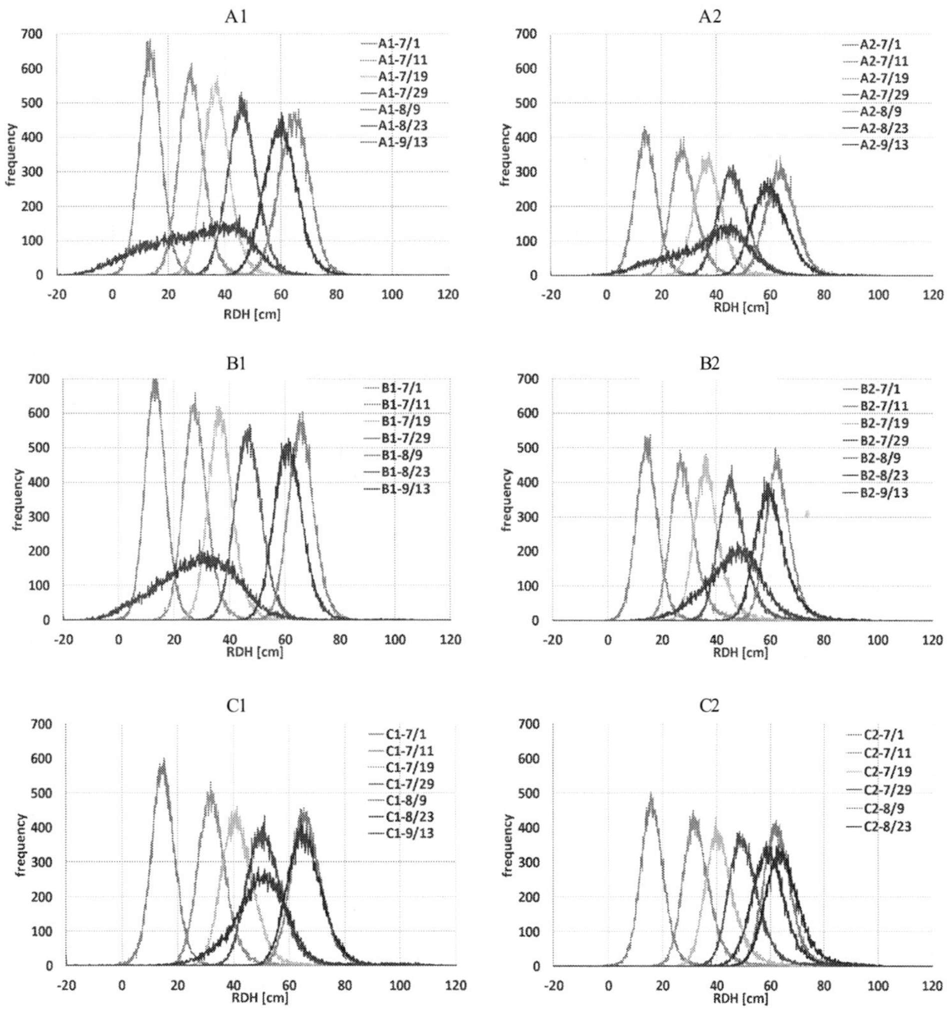

Fig.9 Histogram of each fertilization blocks

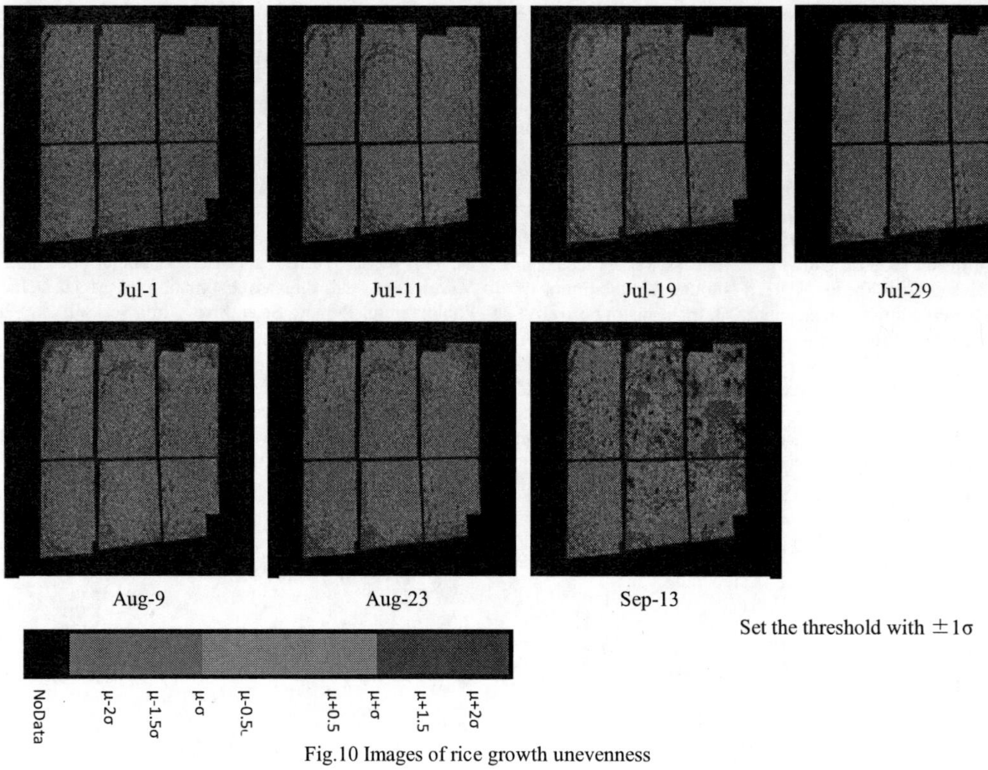

Fig.10 Images of rice growth unevenness

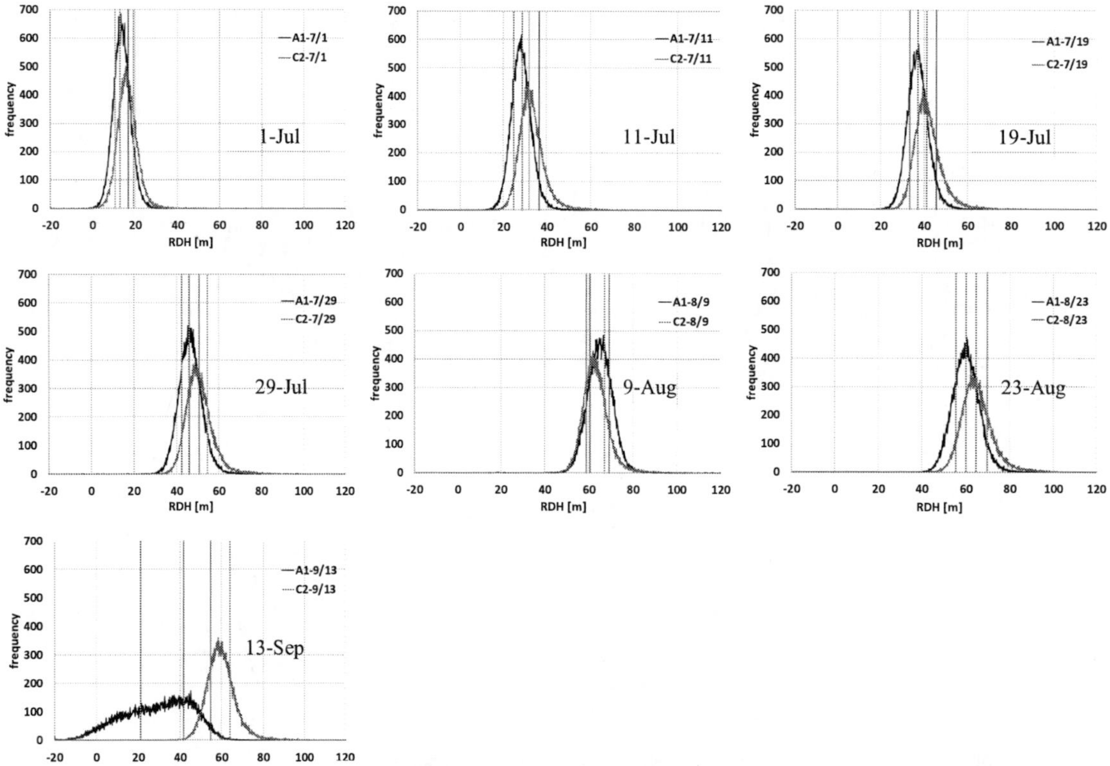

Fig.11 Histogram of A1 and C2 area

https://doi.org/10.5194/isprs-archives-XLII-3-W7-61-2019

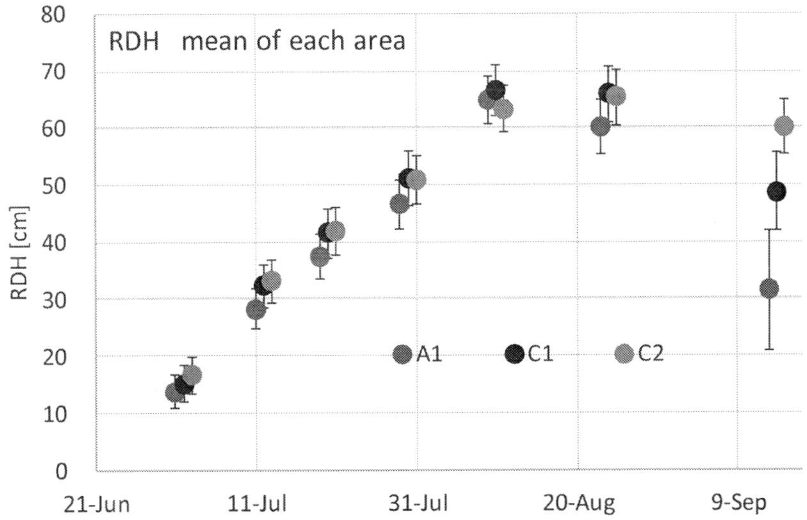

Fig.12 Time-series changing of RDH

SUMMARY OF NATURAL DISASTER RESPONSES BY THE ADVANCED LAND OBSERVING SATELLITE-2 (ALOS-2)

T. Tadono [1, *], M. Ohki [1], T. Abe [1]

[1] Earth Observation Research Center, Japan Aerospace Exploration Agency (JAXA), 2-1-1 Sengen, Tsukuba, Ibaraki, Japan
- (tadono.takeo, ohki.masato, abe.takahiro2)@jaxa.jp

Commission III, WG III/2

KEY WORDS: SAR, L-band, PALSAR-2, disaster analysis, change detection

ABSTRACT:

The Advanced Land Observing Satellite-2 (ALOS-2) was launched on May 24, 2014, and it is operating very well in space more than 4.5 years. The designed mission life is five years as nominal operational phase and the target is over seven years since launch the satellite. The mission objectives of ALOS-2 are 1) disaster monitoring, 2) national land and infrastructure information, 3) cultivated area monitoring, and 4) global forest monitoring. To achieve the objectives, ALOS-2 carries on the Phased Array type L-band Synthetic Aperture Radar-2 (PALSAR-2), which is an active microwave radar using the 1.2 GHz frequency band and observes in day and night times even in bad weather conditions as successor PALSAR instrument onboard ALOS satellite operated from 2006 to 2011. PALSAR-2 instrument has several enhanced features from PALSAR e.g. finer spatial resolution, spotlight observing mode, dual-polarisation ScanSAR. This paper summarises an introduction of typical data analysis results for monitoring natural disasters by ALOS-2 during the operational phase. As the response natural disasters, more than 400 times of the emergency observations have been conducted to identify damages caused by volcanic activities, earthquakes, flooding etc. happened in Japan and the World.

1. INTRODUCTION

Anyone don't know that when and where does a natural disaster happen, however it causes sometime serious damages to human activities and life if it was occurred. When a disaster happens, disaster response agencies would collect information for assessing damages and planning rescue activities as soon as possible. An Earth observation satellite plays an important role in collecting the information. The Advanced Land Observing Satellite-2 (ALOS-2) was launched on May 24, 2014, and it is currently operating very well more than 4.5 years. One of the mission objectives of ALOS-2 is a disaster monitoring of damage areas both in considerable detail and when these areas may be large. The Phased Array type L-band Synthetic Aperture Radar-2 (PALSAR-2) is an active microwave radar instrument and carry on ALOS-2 to contribute the mission objective.
This paper describes examples of data analysis results by ALOS-2 for monitoring natural disasters during the operational phase. As the response natural disasters, more than 400 times of the emergency observations have been conducted to identify damages caused by volcanic activities, earthquakes, flooding etc. occurred in not only Japan but also the World.

2. ALOS-2 AND PALSAR-2 SPECIFICATIONS

Figure 1 shows an in-orbit configuration of ALOS-2, and Table 1 summarizes its specifications (Arikawa *et al.*, 2014). ALOS-2 is a synthetic aperture radar (SAR) mission as a successor PALSAR onboard ALOS satellite that operated from 2006 to 2011 (Shimada *et al.*, 2010). SAR can observe the Earth surface in day- and night-time even under bad weather conditions that effective for monitoring disaster. To share the emergency observations with other satellites effectively, ALOS-2 takes

Figure 1. ALOS-2 in-orbit configuration

Table 1. ALOS-2 specifications

Items		Specifications
Orbit	Type	Sun-synchronous sub-recurrent
	Altitude	628 km at the equator
	LST	12:00 +/- 15 mins
	Revisit	14 days
	Control	< +/- 500 m
Size		3.7 m x 16.5 m x 10.0 m in-orbit
Mass		Approx. 2 tons at launch
Design Life time		Over 5 years (target: 7 years)
Mission data Downlink		X-band: 800 Mbps (16QAM), 400/200 Mbps (QPSK) Ka-band: 278 Mbps (QPSK) *via.* DRTS
Instrument		Phased Array type L-band Synthetic Aperture Radar-2 (PALSAR-2)

* Corresponding author

Table 2. PALSAR-2 observation modes and characteristics

	Spotlight	Ultra Fine	High sensitive	Fine	ScanSAR nominal		ScanSAR wide
Bandwidth	84MHz	84MHz	42MHz	28MHz	14MHz	28MHz	14MHz
Resolution	Rg × Az: 3 × 1m	3m	6m	10m	100m (multilook)		60m (multilook)
Swath	Rg × Az: 25 × 25km	50km	50km	70km	350km (5-scan)		490km (7-scan)
Polarization	SP	SP/DP	SP/DP/QP/CP		SP/DP		
NESZ	-24dB	-24dB	-28dB	-26dB	-26dB	-23dB	-23dB
S/A Rg	25dB	25dB	23dB	25dB	25dB		20dB
S/A Az	20dB	25dB	20dB	23dB	20dB		20dB

SP: HH or VV or HV, DP: HH+HV or VV+VH,
QP: HH+HV+VH+VV, CP: Compact Pol (Experimental)

12:00 of the local sun time (LST), and 14 days of revisit.

Table 2 summarizes PALSAR-2 observation modes and corresponding characteristics (Kankaku *et al.*, 2013). PALSAR-2 provides three major observing modes *i.e.* 1) Spotlight with 3x1 m resolution and 25 km swath width, 2) Stripmap with 3 m (Ultra-Fine), 6 m (High-sensitive), and 10 m (Fine), and 50 to 70 km swath, and 3) ScanSAR with 100 / 60 m resolution and 350 / 490 km swath. Each mode has different polarizations *i.e.* single polarization (SP), dual-pol (DP), and quad-pol (QP). Compact polarimetry (CP) is provided in Stripmap observation mode as experimental purpose. Note that PALSAR-2 can observe both right- and left-looking, and the incidence angle changes from 8 deg. of near-range to 70 deg. of far-range to cover wide observing swath width. This wide coverage is basically important for emergency observation to collect the information as soon as after occurring an event, however analysis of the data acquired by different incidence angles are significantly difficult. An initial operation result of ALOS-2 and PALSAR-2 was summarised by Kankaku *et al.* (2015).

3. EXAMPLES OF DISASTER RESPONSES

Regarding to analyse data by ALOS-2 in disaster responses, it is basically applied the change detection and SAR Interferometry (InSAR) techniques using acquired before and after the event. Therefore, the updating archived data are essential by corresponding observation mode as pre-event data. This is achieved by the "Basic Observation Scenario" (BOS) in ALOS-2, which is a systematic observation strategy and scheduled adequate planning without or minimise conflicts of user requests (Kankaku *et al.*, 2016).

Many large disasters occurred in 2018 in Japan as well as the World. In this section, some typical examples of data analyses using ALOS-2 are introduced.

3.1 Eruption of Mt. Kilauea and Earthquake in Hawaii

Mt. Kilauea in Big Island, Hawaii, US has been erupting since May 3, 2018, and a major earthquake of Mw 6.9 occurred close to Mt. Kilauea on May 4, 2018. Based on the data acquisition request from Jet Propulsion Laboratory (JPL) and United States Geological Survey (USGS), the Japan Aerospace Exploration Agency (JAXA) performed a series of emergency observations by ALOS-2 from May 8, 2018. Since then, intensive emergency observations continuously conducted by several observation modes of ALOS-2 until November 2018.

Figure 2 shows the ScanSAR InSAR interferogram using PALSAR-2 acquired on Jan. 20, 2018 and May 12, 2018. The wider crustal deformation has been revealed. The image shows ~100 cm displacement toward the satellite in south part from Leilani Estates. Moreover, ~50 cm displacement away from satellite was detected around Halemaumau Crater. Figure 3 shows comparisons of amplitude images around Halemaumau

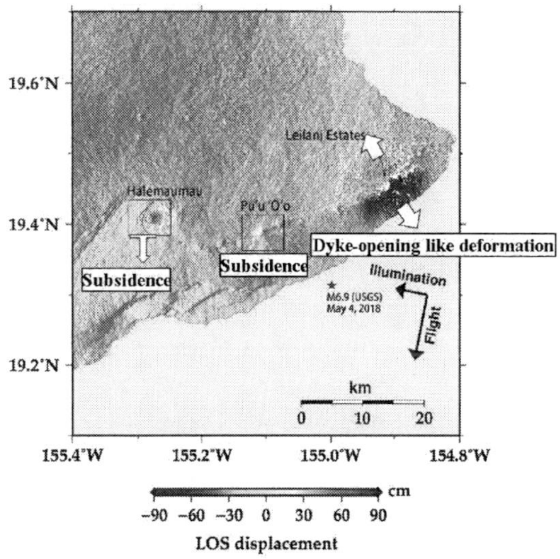

Figure 2. An interferometric image of the PALSAR-2 ScanSAR data acquired before (January 20, 2018) and after (May 12, 2018) the eruption and earthquake

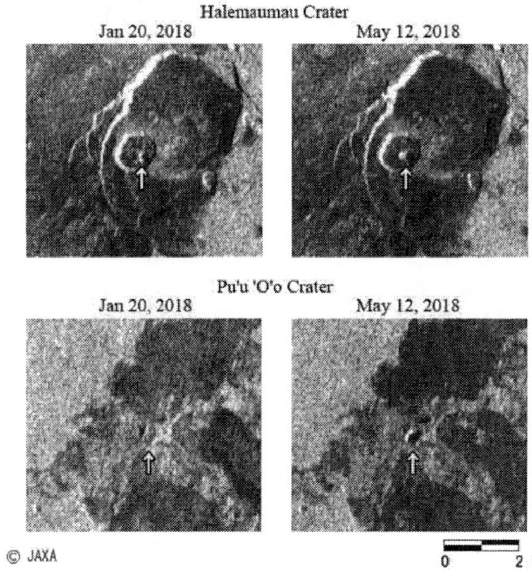

Figure 3. Comparison of the two HH amplitude images acquired before (Jan. 20, 2018) and after (May 12, 2018) the eruption

Figure 4. Color-composite image using Stripmap data acquired before (Feb. 27, 2018) and after (May 22, 2018) assigned as R, G, B=HV_bef, HV_aft, HH_aft. As the result, lava flows were shown in red color.

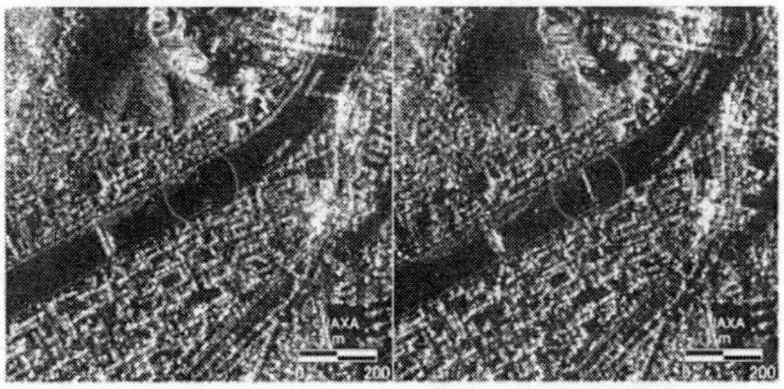

Figure 5. The loss of bridge in Kure City, Hiroshima Pref., Japan from comparison of the two HH amplitude images acquired after (July 8, 2018, left) and before (Apr. 19, 2015, right) due to heavy rain.

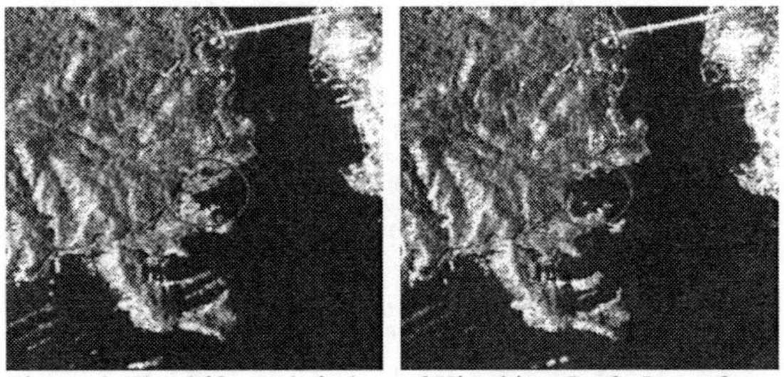

Figure 6. The driftwoods in bay of Hiroshima Pref., Japan from comparison of the two HH amplitude images acquired after (July 21, 2018, left) and before (Apr. 19, 2015, right) due to heavy rain.

crater (upper) and Pu'u 'O'o crater (lower) on May 12 and Jan. 20, 2018. In both craters, terrain inside the craters has changed due to the lowered lava lake level. Figure 4 shows a color-composite image (red: HV polarization on Feb. 27, 2018, green: HV on May 22, 2018, blue: HH on May 22, 2018) around Leilani Estates where eruptive fissures and lava flows can be seen in red color.

These results and information have been shared with JPL, USGS, and USGS Hawaiian Volcano Observatory as soon as after the analysis, and probably used for rescuing activities.

3.2 2018 July heavy rain in western Japan

Based on requests from the Ministry of Land, Infrastructure and Transport (MLIT), Japan, emergency observations by ALOS-2 conducted to identify damaged areas as well as flooding caused by recordable heavy rain mainly in western Japan due to the influence of Typhoon No. 7 in July 2018.

Figures 5 and 6 show damage areas extractions *i.e.* a bridge loss and driftwoods in the bay in Hiroshima Pref., Japan, by comparing two HH polarization images acquired before and after the event. These images acquired by Stripmap Ultra-Fine mode (3 m spatial resolution). Figure 7 shows the expected flooding areas (colored by cyan) in Okayama Pref., Japan using PALSAR-2 acquired after (July 8, 2018) and before (multi-temporal images). The water surface has basically low brightness into SAR amplitude image, however it may contain not only flooding by heavy rain but also seasonal changes of paddy fields etc. To reduce such seasonal change effects and noises in SAR image, multi-temporal images were used here as before the event, which was calculated darker pixels from image acquired same season before (July 2016) and just before the event (Apr. 2018).

(a) The expected flooding around Okayama Pref., Japan

(b) Enlarged image in red square of (a)

Figure 7. The expected flooding areas in Okayama Pref., Japan estimated by PALSAR-2 images acquired after (July 8, 2018) and before (multi-temporal images) colored by cyan.

3.3 Hokkaido Iburi-Toubu earthquake in 2018

A major earthquake of Mw 6.6 occurred in Iburi province, eastern Hokkaido, which is located near Sapporo, Chitose, and Tomakomai Cities at 3:07am on Sep. 6, 2018 (Japan Standard Time) that was caused many landslides in wide ranges of surrounding mountainous areas. Based on the requests from MLIT, Japan, emergency observations by ALOS-2 conducted to identify damaged areas estimation from 11:41am on Sep. 6, 2018.

Figure 8 shows a color-composite image (red: HH on Aug. 23, 2018, green and blue: HH on Sep. 6, 2018) around Iburi province, eastern Hokkaido. In general, forest area has brighter pixel in SAR image, and bare surface darker. Therefore, changing pixels from brighter to darker in forest areas may expect to landslide sites in image acquired before and after the event. As the results, estimated damaged areas by landslides (colored by red) as well as debris areas caused by them (colored by cyan) in Figure 8. These images were acquired by ALOS-2 Stripmap Ultra-Fine mode and 3 m resolution.

(a) The expected landslides around eastern Hokkaido, Japan

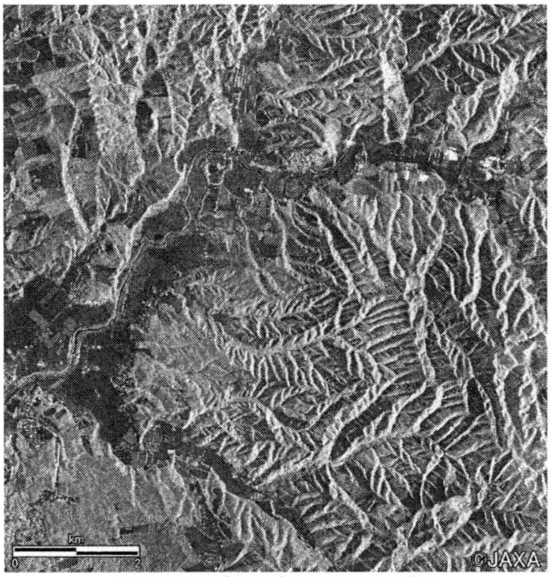

(b) Enlarged image in red square of (a)

Figure 8. The expected landslides (colored by red) and debris areas (colored by cyan) in eastern Hokkaido, Japan estimated by PALSAR-2 images acquired after (Sep. 6, 2018) and before (Aug. 23, 2018) the earthquake.

4. CONCLUSIONS

Recently, Earth observation satellites become an essential tool for the quick responses in natural disaster monitoring. This paper introduced an overview of ALOS-2 and its mission, and examples of practical analysis results for monitoring natural disasters. ALOS-2 is almost completed the nominal operational phase by May 2019 and hopes to operate continuously as longer as possible. During the nominal operational phase, more than 400 times of the emergency observations have been conducted to identify damages caused by volcanic activities, earthquakes, flooding etc. occurred in not only Japan but also the World. The results showed that ALOS-2 has good performances in disaster responses such as all-time observation and high-coherences of L-band SAR, and fine-resolution and wide-swath width observation capabilities. Other examples of data analysis can be found on JAXA EORC (2019).

JAXA is currently under developing of ALOS-3 that is a successor ALOS optical mission with finer resolution and wide swath observation (Tadono *et al.*, 2018), and ALOS-4 that is ALOS-2 follow on SAR mission (Okada *et al.*, 2018). Both missions have an objective to contribute safe and secure social including natural disaster responses, and are planned to be launch around Japanese Fiscal Year 2020.

REFERENCES

Arikawa, Y., Saruwatari, H., Hatooka, Y., and Suzuki, S., 2014. ALOS-2 Launch and Early Orbit Operation Result. In: *Proc. International Geoscience and Remote Sensing Symposium (IGARSS),* IEEE, pp. 3406-3409.

JAXA EORC, 2019. ALOS-2 Research and Application Project, https://www.eorc.jaxa.jp/ALOS/en/index.htm (10 Jan. 2019).

Kankaku, Y., Suzuki, S., and Osawa, Y., 2013. ALOS-2 Mission and Development Status. In: *Proc. International Geoscience and Remote Sensing Symposium (IGARSS),* IEEE, pp. 2396-2399.

Kankaku, Y., Suzuki, S., and Shimada, M., 2015. ALOS-2 First Year Operation Result. In: *Proc. International Geoscience and Remote Sensing Symposium (IGARSS),* IEEE, pp. 4121-4124.

Kankaku, Y., Suzuki, S., Motohka, T., Ohki, M., Natsuaki, R., and Shimada, M., 2016. ALOS-2 Operation Status. In: *Proc. International Geoscience and Remote Sensing Symposium (IGARSS),* IEEE, pp. 3846-3848.

Okada, Y., Yokota, Y., Matsuki, M., Arii, M., and Nakamura, S., 2018. Hardware Performance of PALSAR-3 Onboard ALOS-4. In: *Proc. International Geoscience and Remote Sensing Symposium (IGARSS),* IEEE, pp. 4175-4176.

Shimada, M., T. Tadono, and A. Rosenqvist, 2010. Advanced Land Observing Satellite (ALOS) and Monitoring Global Environmental Change. *Proc. the IEEE,* 98(5), pp. 780-799.

Tadono, T., Oka, A., Watarai, H., Takaku, J., Ohgushi, F., and Doutsu, M., 2018. Image Simulations for the Advanced Optical Satellite (ALOS-3). In: *Proc. International Geoscience and Remote Sensing Symposium (IGARSS),* IEEE, pp. 7731-7734.

https://doi.org/10.5194/isprs-archives-XLII-3-W7-69-2019 | © Authors 2019. CC BY 4.0 License.

The International Archives of the Photogrammetry, Remote Sensing and Spatial Information Sciences, Volume XLII-3/W7, 2019
TC III WG III/2,10 Joint Workshop "Multidisciplinary Remote Sensing for Environmental Monitoring", 12–14 March 2019, Kyoto, Japan

FLOODED AREA EXTRACTION OF RICE PADDY FIELD IN INDONESIA USING SENTINEL-1 SAR DATA

H. Wakabayashi [1], K. Motohashi [1], T. Kitagami [2], B. Tjahjono [3], S. Dewayani[4], D. Hidayat [4], and C. Hongo [2]

[1] College of Engineering, Nihon University, Japan – wakabayashi.hiroyuki@nihon-u.ac.jp
[2] Center for Environmental Remote Sensing, Chiba University, Japan
[3] Bogor Agricultural University, Indonesia
[4] Provincial Office of Food Crops and Horticulture of West Java Province, Indonesia

Commission III, WG III/2

KEY WORDS: flood, backscattering coefficient, linear discriminant analysis, rice paddy, agricultural insurance

ABSTRACT:

The objective of this study is to detect flooded area in rice paddy fields in Indonesia by using remotely sensed data. We used synthetic aperture radar (SAR) data for this purpose, because it is capable of getting high-resolution data in almost all-weather conditions. The paper gives a result of detecting flooded area occurred in our research sites located close to Bandung area, Tegalluar in Bojongsoang district, from the end of February to the beginning of March in 2018. The C-band SAR data acquired by Sentinel-1 were used for this analysis. We derived the gamma-naught threshold for dividing flood and non-flood areas by using a linear discriminant analysis. Discriminant accuracy reached 98% using VV polarization. By using the gamma-naught threshold and rice paddy field mask (GIS data), the rice paddy flooded area could be extracted with good accuracy.

1. INTRODUCTION

Indonesia is one of the leading rice producing countries in the world. Rain plays an important role in rice production especially in the area having clear dry and rainy seasons in the Asian tropical region. On the other hand, the changes in precipitation caused by recent climate change have led to damages in agricultural crops due to natural disasters such as droughts and floods. Since droughts and floods have caused damages in rice paddy fields every year, the trials of the agricultural insurances have started to cover rice paddy damages in Indonesia. So far, the evaluation of the rice paddy damage is carried out by the experts' field survey. However, new methods for the damage evaluations by using remotely sensed data are being investigated in our SATREPS project. This project is titled "development and implementation of new damage assessment process in agricultural insurance as adaptation to climate change for food security". The period of this project is from 2017 to 2022 (Hongo and Baba, 2017).

The objective of this study is to extract flooded area in rice paddy fields in Indonesia by using remotely sensed data. We use synthetic aperture radar (SAR) data for this purpose, because it is capable of getting high-resolution data in almost all-weather conditions. The paper gives a result of extracting flooded area occurred at our research sites located close to Bandung area from the end of February to the beginning of March in 2018. The C-band SAR data acquired by Sentinel-1 is used for extracting flooded area in this analysis. The UAV data as well as the ground photos were used for verifying the extracted result.

2. TEST SITE AND SAR DATA

2.1 Test site

Our test site in this research is Bojongsoang district that is located about 4 km south east of Bandung City in West Java province. The south of Bojongsoang is surrounded by Citarum River and flooding at paddy fields occurs every year in rainy season from December to March. The flood was occurred in the rice paddy fields in Tegalluar area located in the east part of Bojongsoang during the period from the end of February to the beginning of March in 2018. Figure 1 shows a land use map in Bojongsoang created by the Bogor Agricultural University (Hanidya, 2018). We can recognize that 63 % of land area is used as rice paddy fields followed by settlements, buildings and residential areas. Figure 2 gives a series of false-coloured images covering Tegalluar area taken by PLANET satellites during the period from Feb. 21 to Mar. 4 in 2018. The interval of the images was 2 days and we could recognize the flooded area in each image. The flooding was occurred on Feb. 21 in the center of Tegalluar and expanded till Feb. 27, which reached the maximum flooded area. The area of flood decreased after March.

2.2 SAR data

We used Sentinel-1 C-band SAR data for the detection of flooded area in the rice paddy fields, because it is capable of getting relatively high-resolution data in almost all-weather conditions. The characteristics of the Sentinel-1's C-band SAR data are given in Table 1. The SAR data used in this research were acquired in the Interferometric Wide (IW) swath mode. This mode employs the Terrain Observation with Progressive Scans SAR (TOPSAR) technique to acquire data over a wider area using three sub-swaths. The IW mode acquires data over a 250-km swath with a 5-meter (range) by 20-meter (azimuth) spatial resolution. The incidence angle of IW mode is from 29 to 46°. The incidence angle at our test site was about 43 degrees. The VV- and VH- polarization data were used in this study. Level-1 Ground Range Detected (GRD) data were downloaded from the European Space Agency (ESA) Copernicus Open Access Hub and input into the Sentinel Toolbox software (ESA, 2018) for further analysis. The SAR data were firstly calibrated to backscattering coefficients and then transformed onto the UTM coordinates. The pixel spacing on the UTM coordinates was set to 10-meter by 10-meter. Corrections for the foreshortening as well as the local incidence angle (Terrain Flattening) were applied, because the analysed areas were located in the average altitude of 700m and the local incidence angle at the observation target was varied in wide swath mode.

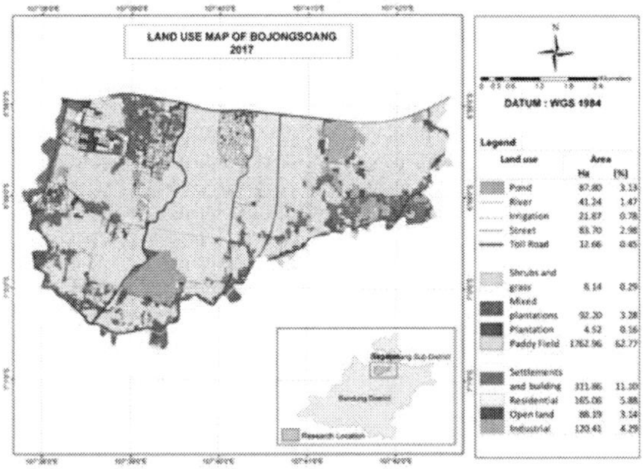

Figure 1. Land use map of Bojongsoang created by Bogor Agricultural University (IPB).

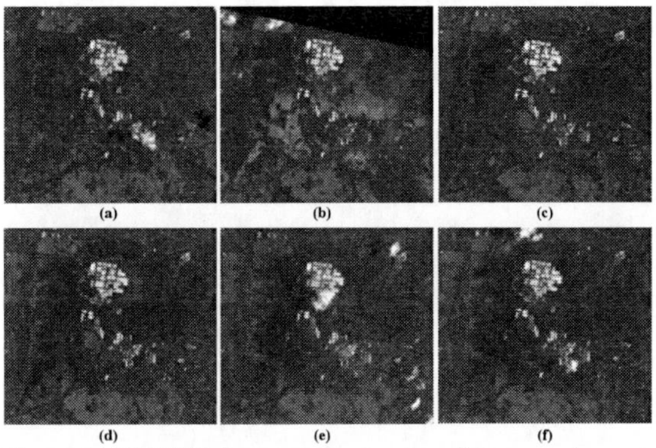

Figure 2. PLANET images covering Tegalluar area in east Bojongsoang acquired from Feb. 21 to Mar. 4 in 2018. (a) Feb.21, 2018, (b) Feb.23, 2018, (c) Feb.25, 2018, (d) Feb.27, 2018, (e) Mar.02, 2018, and (f) Mar.04, 2018.

Table 1. Sentinel-1's C-band SAR characteristics

	Sentinel-1A/B
Center frequency	5.405 GHz(C-band)
Bandwidth	10 - 100 MHz
Altitude	693 km
Image width	80-410 km
Range resolution	5 m
Azimuth resolution	20 m
NE sigma zero	-22 dB
Incidence angle	20-47 deg.
Polarimetry	HH+HV or VV+VH
Tx power	4.4 KW

3. SAR DATA ANALYSIS FOR FLOOD DETECTION

3.1 SAR data pre-processing

The Sentinel-1's C-band SAR data covering our test site were used in this analysis, which were acquired twice on Feb. 17 and Mar. 1 in 2018. The former data were acquired in non-flooded situation and the latter data were in flooded situation. The SRTM-1 30-meter DEM was used for the corrections of foreshortening as well as the local incidence angle in order to reduce incidence angle distortions on the SAR data. Our procedure to process SAR data using Sentinel Toolbox are given as follows,

(1) Apply Orbit data
(2) Calibrate SAR data to get Beta-naught
(3) Local incidence angle correction to get gamma-naught using SRTM-1 DEM (Radiometric Terrain Flattening)
(4) Foreshortening correction using SRTM-1 DEM (Range-Doppler Terrain Correction)
(5) Transform onto UTM coordinates

3.2 Flooded area detection

Figure 3 shows the change of backscattering mechanisms for rice paddy fields as the water depth increase. As shown in this figure, the SAR backscattering coefficients in both VV and VH polarizations decrease, when the flood occurs in rice paddy fields. Since the thresholds of backscattering coefficient to divide flood and non-flood areas were successfully applied to the SAR data in various frequencies (Henry et al. 2006, Martinis et al. 2015, and Ohki et al. 2016), this method can be used for the Sentinel-1's C-band SAR data as well.

We quantitatively verified classification accuracy of flood and non-flood areas using multivariate discriminant analysis (Fisher, 1936 and Otsu, 1979). We first determined within-class and between-class variance of two classes, flood and non-flood areas, which were recognized in SAR data acquired on February 17 and March 1. We applied a linear discriminant analysis (LDA) in order to separate flood and non-flood areas by gamma-naught, ultimately determining a threshold separating the two classes. Through LDA, a separation line between two classes can be established.

In order to determine a threshold to divide flood and non-flood areas, the gamma-naught in 30 small sampling points were extracted in both flood (March 1) and non-flood (February 17) data. Figure 4(a) shows the 30 sampling points in Tegalluar area. We extracted 540 samples in total, because each point in Figure 4 included 3 by 3 samples. Figure 4(b) and 4(c) give the histograms of VV and VH gamma-naught backscattering for both flood and non-flood areas, and Table 2 shows the result of LDA for determining threshold values. From the correlation ratio and the discriminant accuracy in this table, we can conclude that VV data are superior to VH data in discriminating flooded area in the rice paddy fields.

The median filter was applied to the gamma-naught image with 3 by 3 pixels moving window in order to reduce speckle noise. Then, the rice paddy field in Figure 1 (Figure 5(a)) is used for masking the gamma-naught images as given in Figure 5(b) and 5(c). After applying the thresholds in Table 2 to the gamma-naught images in order to detect flooded area, the differential binary image between flood and non-flood images is shown in Figure 6. The black area in this figure indicates the rice paddy fields where the backscatter was decreased from Feb.17 to Mar.1.

Figure 7(a) gives the ortho-mosaic image acquired by the UAV on Feb. 27, 2018. This image was created from the images taken by the near-infrared camera mounted on the UAV with 150 m altitude. It covers about 80 ha of southwest Tegalluar area. The colorized image was assigned near-infrared to Red, and the flooded area is easily recognized in this image. Figure 7(b) and 7(c) show examples of ground photos taken by the pest observer (damage assessor) in Bojongsoang. The extracted flood area can be verified by the image and photos in Figure 7.

This contribution has been peer-reviewed.
https://doi.org/10.5194/isprs-archives-XLII-3-W7-73-2019 | © Authors 2019. CC BY 4.0 License.

3.3 Texture analysis

We tried to estimate flooded area by using the method other than backscattering based method. We used in particular the Gray-Level Co-occurrence Matrix (GLCM) texture analysis (Haralick et al. 1973) for detecting flooded area. When the flood occurred in rice paddy fields, the backscattering is dominated by specular reflections. Since the variance of texture decreases, the spatial entropy may also decrease.

Figure 8(a) and 8(b) give examples of spatial entropy calculated based on GLCM, in which the number of level is 32 with 5 by 5 window calculating texture at the angle of 0 deg. Since the spatial entropy decreases at the flooded area, we may use this method to find the flooded area with SAR data.

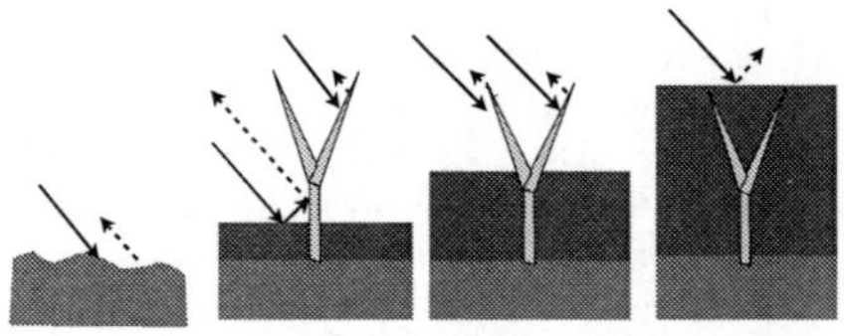

Paddy status	Before planting	Irrigation	Increase water level	Completely flood
Dominant scattering	Surface scattering from the ground	Volume scattering+ Double bounce	Volume scattering	Specular scattering
Change of backscattring		increase	decrease	decrease

Figure 3. Change of backscattering mechanisms of rice paddy fields.

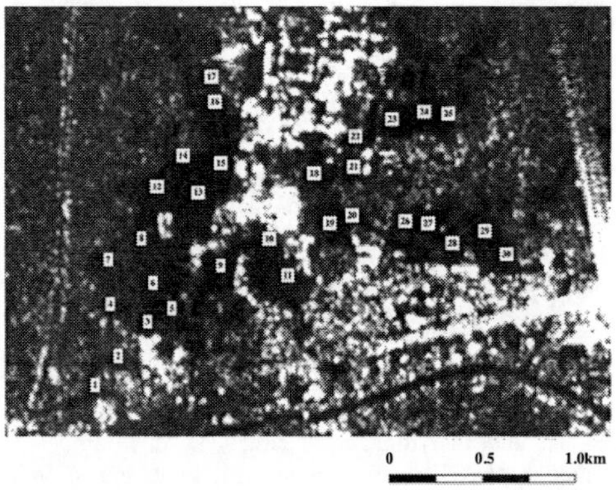

(a)

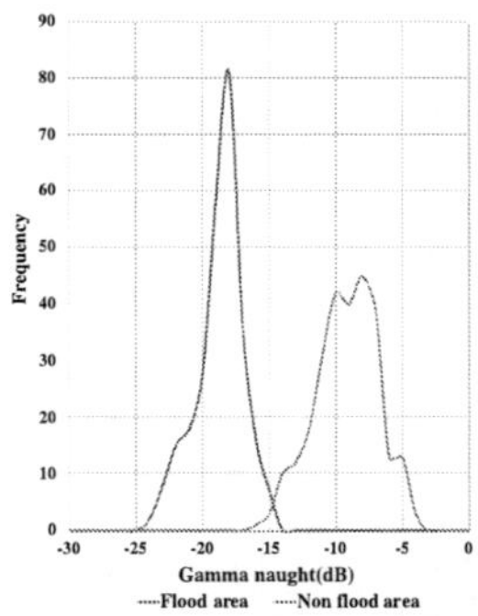

(b)

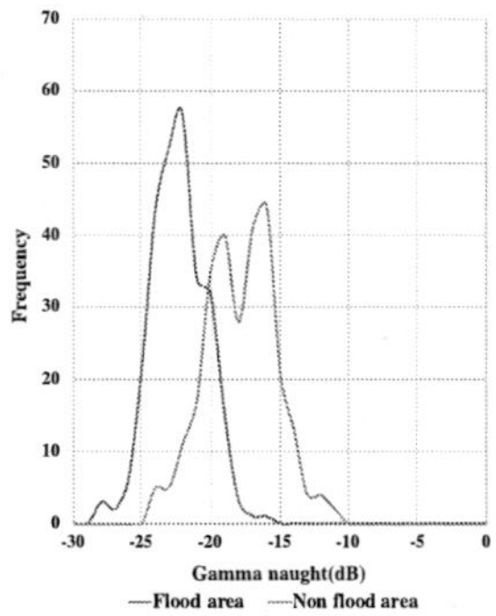

(c)

Figure 4. Gamma-naught distributions for both flood and non-flood rice paddy fields. (a) 30 sampling points in Tegalluar area, (b) VV gamma-naught and (c) VH gamma-naught distributions in both flood and non-flood rice paddy fields.

Table 2. Result of linear discriminant analysis for determining gamma-naught thresholds.

Polarization	Threshold(dB)	Correlation ratio	Discriminant accuracy
VV	-14.0	0.84	0.98
VH	-20.1	0.49	0.83

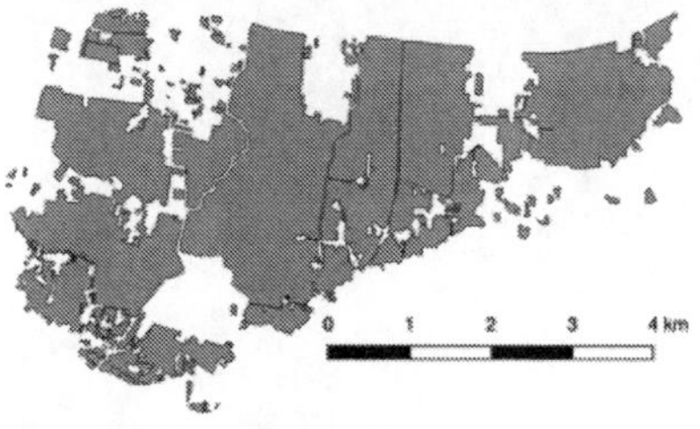

(a)

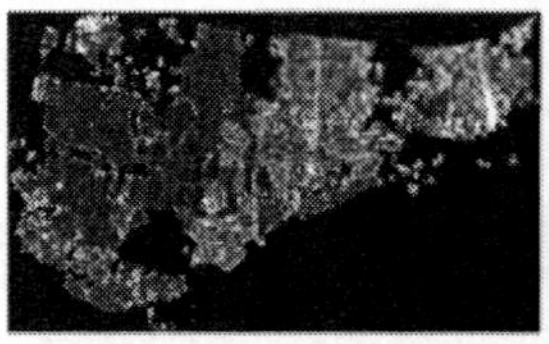

(b)

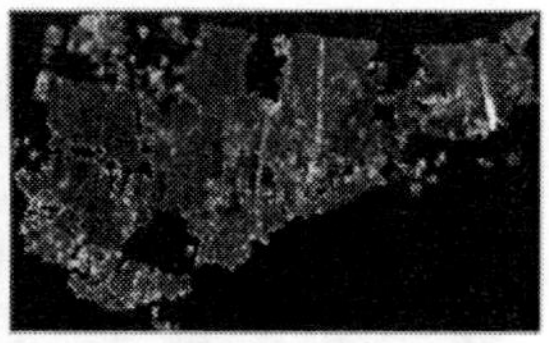

(c)

Figure 5. Figure 6. Gamma-naught images after applying rice paddy field mask. (a) rice paddy field mask, (b) Feb.17,2018, and (c) Mar.1, 2018.

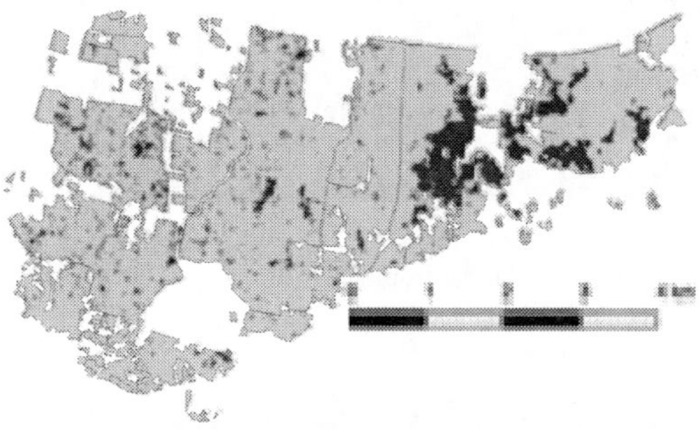

Figure 6. Extracted flooded area from SAR data. Black area indicates the rice paddy fields where the backscattering is decreased from Feb.17 to Mar.1.

4. SUMMARY

Agricultural incurrence has been implemented in Indonesia for three years, and the damage assessment of rice paddy is conducted by in-situ observations. We are now investigating new methods to evaluate the damage in rice paddy fields by using remotely sensed data in our SATREPS project. In this paper, we tried to evaluate damage in rice paddy field caused by the flood in Bojongsoang from the end of February to the beginning of March in 2018.

We can conclude that the detection of flood area is possible by using the threshold of C-band SAR gamma-naught with good accuracies. The ortho-mosaic UAV image as well as ground photos taken by pest observer verify the estimated flood area. Since the knowledge of rice paddy growing stage is necessary for accurate damage assessment caused by flooding, we will conduct the further investigation to monitor rice paddy growing stage (Nguyen et al. 2016) combined with our flood detection method.

https://doi.org/10.5194/isprs-archives-XLII-3-W7-73-2019 | © Authors 2019. CC BY 4.0 License.

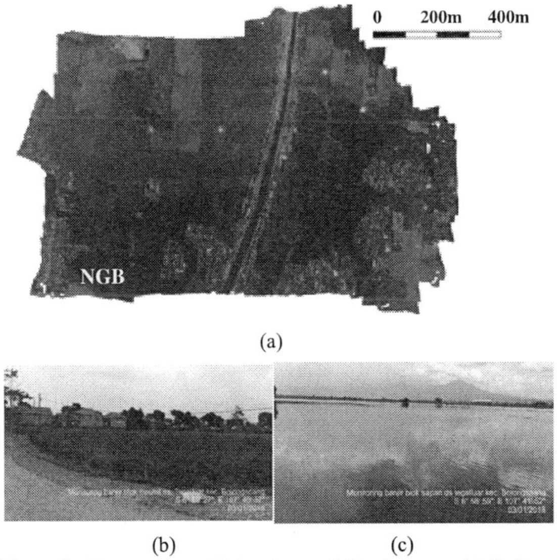

Figure 7. Dataset for verifying detected flooded area. (a) Ortho-mosaic image covering 80-ha of Tegalluar area in Bojongsoang, which was acquired by the UAV mounting the near-infrared camera on Feb.27. (b) and (c) Examples of ground photos taken by the pest observer in Bojongsoang.

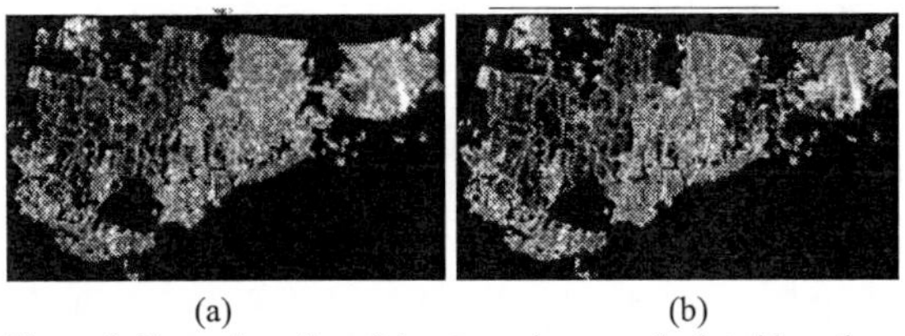

Figure 8. Examples of spatial entropy image calculated based on GLCM. (a) Feb.17, 2018, and (b) Mar.1, 2018.

ACKNOWLEDGEMENTS

This work was conducted under the SATREPS project, "Development and Implementation of New Damage Assessment Process in Agricultural Insurance as Adaptation to Climate Change for Food Security". The Sentinel-1 data were provided by the ESA. This research has been partly conducted by the cooperative research with Center for Environmental Remote Sensing, Chiba University. We are grateful to the pest observers for their fieldwork at Bojongsoang. We also thank Sella Atika Sri Kumala, Farah Satira Hanidya, and Yuki Watanabe for their works making and analyzing GIS data in Bojongsoang.

REFERENCES

Hongo, C, and Baba, B.,2017. SATREPS Project. https://www.jst.go.jp/global/english/kadai/h2804_indonesia.html

Hanidya, F. S., 2018. Flood Vulnerability Analysis in Bojongsoang District, Bandung Regency, Graduation Research of Land Resource Management Study Program, Bogor Agricultural University (IPB).

ESA Sentinel toolbox develop team: https://sentinel.esa.int/web/sentinel/toolboxes

Henry, J. B., Chastanet, P., Fellah, K. ,and Desnos ,Y. L., 2006. Envisat multi-polarized ASAR data for flood mapping, International Journal of Remote Sensing, Vol. 27, No. 10, pp.1921–1929.

Martinis, S., Kersten, J., and Twele, A., 2015. A fully automated TerraSAR-X based flood service, ISPRS Journal of Photogrammetry and Remote Sensing 104, pp.203–212.

Ohki, M., Watanabe, M., Natsuaki, R., Motohka, T., Nagai, H., Tadono, T., Suzuki, S., Ishii, K., Itoh, T., Yamanoguchi, T. and Shimada, M., 2016. Flood Area Detection Using ALOS-2 PALSAR-2 Data for the 2015 Heavy Rainfall Disaster in the Kanto and Tohoku Area, Japan, Journal of The Remote Sensing Society of Japan, Vol.36, No.4, pp.348-359.

Fisher, R.A., 1936. Annals of Eugenics 7(2) : The use of multiple measurements in taxonomic problems, lackwell Publishing Ltd, University College London, pp. 179-188.

Otsu, N., 1979. A Threshold Selection Method from Gray-Level Histogram, IEEE Trans. on System Man Cybernetics, SMC, Vol,9(I), pp. 62-66.

Haralick, R.M., Shanmugam, K., and Dinstein, I., 1973. Texture Features for Image Classification, IEEE Trans. on Systems, Man and Cybernetics, Vo. 3, No. 6, pp. 610-621.

Nguyen, D. B., Gruber, A., and Wagner, W., 2016. Mapping rice extent and cropping scheme in the Mekong Delta using Sentinel-1A data, Remote Sensing Letters, Vol. 7, No.12, pp. 1209-1218.

CHLOROPHYLL ESTIMATION OF LAKE WATER AND COASTAL WATER USING LANDSAT-8 AND SENTINEL-2A SATELLITE

S.Yadav [1]*, Y.Yamashiki [1], J.Susaki [2], Y.Yamashita [3], K.Ishikawa [4]

[1] Graduate School of Advanced Integrated Studies on Human Survivability, Kyoto University, Kyoto 606-8501, Japan - shwtdv@gmail.com *, yamashiki.yosuke.3u@kyoto-u.ac.jp
[2] Department of Civil and Earth Resources Engineering, Graduate School of Engineering, Kyoto University, Kyoto 615-8540, Japan - susaki.junichi.3r@kyoto-u.ac.jp
[3]Kyoto University, Field Science Education and Research Center, Nagahama 625-0086, Japan - yoh@kais.kyoto-u.ac.jp
[4] Lake Biwa Environmental Research Institute (LBERI), Otsu 520-0022, Japan - ishikawa-k@lberi.jp

Commission III, WG III/10

KEY WORDS: Chlorophyll-a, Landsat-8, Sentinel-2A, Coastal Water, Freshwater, Satellite-Remote Sensing

ABSTRACT:

Chlorophyll-a is an optically active compound (OAC) commonly used as a proxy for phytoplankton biomass in an aquatic environment. Retrieving the concentration of chlorophyll-a remains a challenge due to the presence of several OAC particularly in water bodies which are in proximity to the land-based activities. In this study, an effort has been made to estimate the chlorophyll-a concentration of both the freshwater Lake Biwa and the coastal water of Wakasa Bay in Japan. A spectral decomposition algorithm was used to determine the chlorophyll-a using the satellite images. The algorithm was applied to the satellite images from two different sensors namely Landsat-8 /OLI and Sentinel-2A/MSI. The satellite-derived chlorophyll-a concentration for the lake and coastal water from two different sensors were compared to assess the performance of both the sensors. The accuracy of the chlorophyll-a results derived from the images was evaluated with the in-situ measurement data of the chlorophyll-a for the Lake Biwa and the coastal water of Wakasa Bay. Both satellite sensors appear to give the best results for the coastal water ($R^2 > 0.80$) with an RMSE < 0.3 µg/L. However, slight underestimation of chlorophyll-a noted for the Landsat-8 image with an increase in chlorophyll-a concentration. For the lake water, Sentinel-2A results were relatively better ($R^2 > 0.70$) than Landsat-8, with an RMSE of < 1.0 µg/L. The obtained results will be useful to evaluate the primary productivity of both freshwater and coastal water body.

1. INTRODUCTION

1.1 Background

At large, global water bodies are classified into two, Case-1 and Case-2 water types. In the case-1 water, phytoplankton dominates the inherent optical properties (IOPs) of water (e.g., open ocean water). While Case-2 water are the complex water bodies influenced by variety of optically active compound (OAC) such as phytoplankton, colored dissolved organic matter (CDOM), inorganic particles and other suspended matters (e.g., coastal and inland waters) (Morel, 1988; Mobley et al., 2004; Matsushita et al., 2012). In the recent decade, the increased pollutant load degraded the water quality of case-2 water mainly rivers, lakes and coastal oceans (Carpenter et al., 1998; Kändler et al., 2017; Zhang et al., 2018; Bhagowati and Ahamad, 2018). Land-based activities adjacent to the coastal and inland water bodies directly affect the water quality and aquatic ecosystem through numerous point sources (i.e., wastewater effluent, runoff from waste storages, stormwater outfalls) and non-pint sources of pollution (i.e., atmospheric deposition, agricultural runoff, deforestation and land conversion) (Fredston-Hermann et al. 2016; Kandler et al. 2017). Riverine transport is the primary mechanism for the direct impact of the terrestrial-human activities on the nearshore coastal waters and estuaries (Fredston-Hermann et al. 2016).

Excess sediment and nutrient transport from anthropogenic sources (e.g., agriculture, industries and sewage disposal) in case-2 water often lead to increased turbidity and reduced light availability, which in turn negatively impacts the biogenic species in the aquatic environment such as fishes and corals in freshwater and coastal water (Silva et al., 2016; Fredston-Hermann et al., 2016; Lintern et al., 2017). Nutrient load to inland waters (Le et al., 2009; Yadav et al., 2017) and coastal zones (Cloern and Jassby, 2010) lead to eutrophication including harmful algal bloom. A most serious threat from eutrophication is hypoxia which is the decrease in the dissolved oxygen (DO) levels in bottom waters, created as planktonic algae decompose. Hypoxia occurs when DO falls below 2ml of O_2/liter, thus increased mortality of the benthic fauna (Diaz, 2009; Shulkin et al. 2012). Chlorophyll-a (Chl-a) an optically active compound (OAC) and a proxy for phytoplankton biomass in an aquatic environment (Cullen, 1982; Blondeau-Patissier et al., 2014), is frequently used as a key indicator to describe the water quality, pollution effect and biophysical status of the water bodies (Carpenter et al., 1998; Shi et al. 2013; Cheng et al. 2013; Wozniak et al., 2014). Chlorophyll-a is the most common photosynthetically active pigment present in almost all phytoplankton species, which influences the optical properties of the freshwater and seawater by increasing the light backscattering with spectrally localized water-leaving radiance (Blondeau-Patissier et al., 2014).
Remote sensing techniques which provide near-real-time synoptic overview of the large areas commonly used chlorophyll-a for monitoring the phytoplankton abundance in both inland water (Le et al., 2009; Cheng et al., 2013; Shi et al., 2013) and coastal waters (Reinart and Kutser, 2006; Le et al., 2013;

* S.Yadav

Blondeau-Patissier et al., 2014; Wozniak et al., 2014). Algorithms that rely mostly on the detection of specific spectral features using blue, green, yellow, red or near-infrared (NIR) portion of the reflectance is well suited for discrimination of algal blooms from other naturally occurring phenomena. However, the reliability of the measured reflectance is often hampered by its sensitivity to such as the thickness of the floating algal layer, suspended particulates, bottom reflectance (in shallow water) and atmospheric corrections. Thus a satellite-based technique along with the knowledge of the study area aid in the accurate estimation of the phytoplankton abundance (Reinart and Kutser, 2006; Blondeau-Patissier et al., 2014).

Retrieving the concentration of chlorophyll-a using satellite-remote sensing remains a challenge not only due to the presence of several OAC but also of the available sensors specifications (Ha et al., 2017), particularly in water bodies which are in proximity to the land-based activities. Some studies showed that the broad wavelength spectral data available on current satellites (i.e., Landsat, SPOT) does not allow the estimation of chlorophyll-a in turbid water (Dekker and Peters, 1993; Ha et al., 2017). On the other hand, Sentinel-2A MSI with high spectral (13 bands) resolution considered highly suitable for monitoring chlorophyll-a in coastal and inland waters due to its red-edge band near 705 nm (band 5) and second peak of chlorophyll absorption in red band (665 nm, band 4) (Toming et al., 2016; Ha et al., 2017). In this study, an effort has been made to estimate the chlorophyll-a concentration of both the freshwater Lake Biwa and the coastal water of Wakasa Bay in Japan and the feasibility of two different sensors mainly Landsat-8 OLI and Sentinel-2A MSI were assessed and compared for both the water bodies. Furthermore, the applicability of the spectral decomposition technique was tested on both coastal and inland water body.

1.2 Study Area

The coastal water of Wakasa Bay (mainly Yura Estuary) and the Lake Biwa which is the largest freshwater lake in Japan was studied in this paper.

Lake Biwa (35°20′59.48″N and 136°10′33.16″E) is located in the Honshu Island, southwest of Japan (Figure 1). The lake is an important freshwater source for over 14 million people in the Kansai region of Japan. The surface area of the lake is 670 km^2 and a total catchment area of 3838 km^2. The lake encompasses two basins, the north basin (618 km^2) which is mesotrophic and deep (the mean and the maximum depths are 44 m and 104 m, respectively), and the south basin (52 km^2) which is eutrophic and shallow (the mean and the maximum depths are 3.5 m and 8 m, respectively) (Yadav et al., 2017). Lake Biwa experienced eutrophication (mainly the south basin) during the economic growth period after the 1960s. The land use change in the catchment area is primarily attributed to the ongoing urbanization and change in the industrial structure in the periphery of the south basin, contributing to eutrophication. To curb the eutrophication and to restore the lake ecosystem, the Shiga Prefectural Government enacted the ordinance Prevention of Eutrophication of Lake Biwa in 1979 followed by the formulation of the Water Quality Conservation Plan in 1986. Consequently, water quality of the lake improved which led to an upsurge in the growth of submerged Macrophytes in the lake (Ishikawa et al., 2002; Yadav et al., 2017).
On the other hand, Wakasa Bay has a wide opening to the Japan Sea (35°31′51.5″ N and 135°17′29.3″ E). Yura river which is one of the major river of the main island of Japan flows into the western part of the Wakasa Bay (Figure 1), also known as Tango sea. The river is 146 km in length and has a total catchment area of 1880 km^2 with an annual discharge of 41 m^3 s^{-1}. The river discharge is high in the winter due to snow melting and low in summer. Besides, the area is also subjected to cyclonic rain depression in summer (Kasai et al., 2010). The Bay area generally has a steep topography near the continental shelf, except on the coast of Yura River where the sea bed inclination is gentle owing to the sediment deposition from the river. Coarse sandy bottom characterizes the shallow coastal area (3-5 m) and fine sand at the coastal area of 10-20 m (Antonio et al., 2010). The Yura estuary is also classified as the microtidal estuary as the typical tidal range in the Yura estuary is < 0.5 m (Kasai et al., 2010).

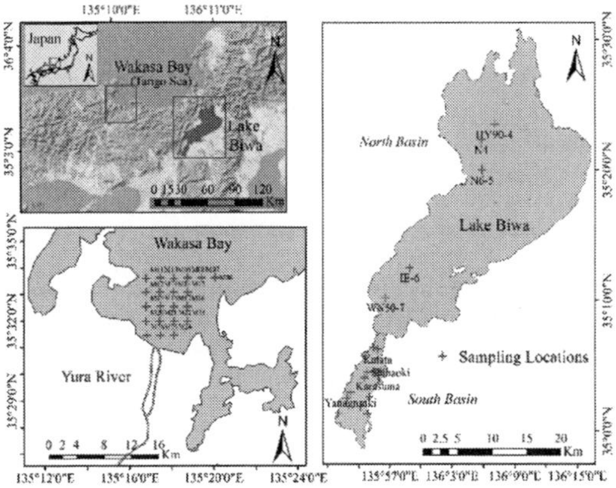

Figure 1. Geographical location of the freshwater lake (Lake Biwa) and coastal water (Wakasa Bay) in Japan.

2. MATERIALS AND METHODS

2.1 Field Data

The vertical profile of the chlorophyll-a, water temperature and many other parameters such as DO and turbidity were measured using the calibrated fluorometer (Chlorotech, ACL200 PDK, Alec Electronics) for Lake Biwa. The measurements were conducted together with Lake Biwa Environment Research Center (LBERI) between 31 July 2018 to 2 August 2018. Chlorophyll-a was measured for 18 locations. Water samples were collected from 0.3 m from the water surface for further measurements in the laboratory including phytoplankton species classification which is beyond the scope of this paper.

A vertical profile of the chlorophyll-a, salinity, water temperature and DO was measured at 21 locations in the Wakasa Bay (Yura Estuary) using CTD profiler (Compact CTD, Alec Electronics) and calibrated Rinko Chlorophyll Fluorometer. The water samples were taken from each location at 0.2 - 0.3 m from the surface. The filtered water samples (precombusted GF/F filters) were used for chlorophyll-a extraction using 90% acetone (12 h) which was then measured using the calibrated fluorometer. The data from the CTD profiler was then calibrated using the measured chlorophyll-a. The observations were carried out at every 1km from the mouth of Yura Rivers. The spectral data at each location were measured using the ASD FieldSpec^{TM} Pro JR Field Portable Spectroradiometer (350 – 2500 nm) and Handheld spectroradiometer (400 -1100 nm).

In addition, the reflectance characteristic of the aquatic vegetation species samples obtained by the diver survey were measured in the laboratory ASD FieldSpec™ Pro JR Field Portable Spectroradiometer and Handheld spectroradiometer (Yadav et al., 2017). Samples collected particularly from the south basin of Lake Biwa and the shorelines of the north basin, whereas in the Wakasa Bay samples collected from the two locations near the shallow coast of Yura River (3-5 m).

2.2 Satellite Remote Sensing Data

The Landsat-8 OLI image (30 m spatial resolution) and Sentinel-2A MSI image (10, 20, 60 m spatial resolution) for Wakasa Bay and Lake Biwa were acquired U.S. Geological Survey (USGS) EarthExplorer (USGS, https://earthexplorer.usgs.gov/.). The satellite images were downloaded based on survey time, geographical extent and environmental conditions. The Landsat-8 acquisition date was 3 August 2018 and for Sentinel-2A acquisition date was 4 August 2018. Image rectification and geoprocessing were achieved using ENVI 5.2 image analysis software and ArcMap 10.3.1. Apart from rescaling to top of atmosphere (TOA) reflectance, the images were atmospherically corrected using the mid-latitude summer atmospheric model and maritime aerosol model in the atmospheric correction module FLAASH of ENVI 5.2. Field measured spectral data was further used to validate the atmospheric correction of the images.

2.3 Chlorophyll-a Estimation Algorithm

Spectral Decomposition Algorithm was used to estimate the chlorophyll-a in the mainly Lake Biwa and Wakasa Bay. In this algorithm, the mixed reflectance spectra of a given pixel are conceptualized as a linear combination of potential endmembers substantially contributing to the pixel reflectance (Yadav et al., 2017). We assume that this algorithm which was previously being used only for the shallow lakes can be applied to other optically-complex case-2 water bodies influenced by sediment and nutrient loads which also supports the growth of aquatic vegetation. The mixed-pixel reflectance in the Spectral Decomposition Algorithm is expressed as in Equation1:

$$R(\lambda) = a_p \times R_p(\lambda) + a_n \times R_n(\lambda) + a_v \times R_v(\lambda) + a_w \times R_W(\lambda) \quad (1)$$

Where, a_p, a_n, a_v and a_w symbolize the decomposition coefficients of phytoplankton, non- phytoplankton suspended solids (NPSS), submerged aquatic vegetation (SAV), and water, respectively. R_p, R_n, R_v and R_w represent the standard reflectance spectra of phytoplankton, NPSS, SAV, and water, respectively. The wavelength is expressed by λ in the equation.

Four decomposition coefficients were estimated by applying Equation (1) to each of the Landsat-8 bands, blue, green, red and NIR (i.e., band 2 to band 5). Similar spectral bands were used for Sentinel-2A image (i.e., band 2, 3, 4 and 8A) representing blue green, red and NIR. In Sentinel-2A, band 8A has 20 m spatial resolution comparison to blue, green, red band with 10 m spatial resolution. Therefore, band 8A of Sentinel-2A was resampled to 10 m resolution in ArcGIS before using it in the equation. In addition, the standard reflectance spectra of the endmembers were simulated using the Bio-Optical model (Yadav et al., 2017), shown in Equation 2. For aquatic vegetation standard reflectance was selected from the average reflectance of the mixture of the species measured in the laboratory. The vegetation cover assumed to be homogeneous. The details of the equation can be found in (Lee et al., 1998; Oyama et al., 2009).

$$\begin{aligned} r_{rs} = {} & (0.070 + 0.16u^{0.752})\, u \Big(1 \\ & - 1.03 \\ & \times Exp\Big\{-\Big[\frac{1}{\cos\theta_W} \\ & + 1.2(1 + 2.0u)^{0.5}\Big](a \\ & + b_b)H\Big\}\Big) \\ & + 0.31\rho \\ & \times Exp\Big\{-\Big[\frac{1}{\cos\theta_W} \\ & + 1.1(1 + 4.9u)^{0.5}\Big](a \\ & + b_b)H\Big\} \end{aligned} \quad (2)$$

where r_{rs} is the subsurface reflectance, a and b_b are the total absorption and backscattering coefficient, respectively, contributed by the pure water, phytoplankton, NPSS, and cDOM in the water column. In addition, θ_W indicates the subsurface solar zenith angle, H indicates the bottom depth (m), and r indicates the remote sensing reflectance of the shallow water. Whereas, u can be expressed by a and b_b.

Furthermore, the phytoplankton decomposition coefficient a_p obtained by solving the four bands using Spectra Decomposition Algorithm was used as an independent variable to estimate the phytoplankton biomass. The regression model was developed using the a_p and the natural logarithm of Chl-a (ln(Chl-a))for both Landsat-8 and Sentinel-2A images as given in Equation 3:

$$\ln(Chla) = A + B \times a_p \quad (3)$$

Where, A and B represents the regression coefficients in the equation. The chlorophyll-a was estimated for both Wakasa Bay and Lake Biwa using the above algorithm on Landsat-8 and Sentinel-2A images. The results were compared with the in-situ measured data of chlorophyll-a for both the water bodies.

3. RESULTS

Chlorophyll-a for Lake Biwa and Wakasa Bay was determined by applying the algorithm to both Landsat -8 and Sentinel-2A image. The satellite-derived chlorophyll-a of both the sensors were compared and evaluated using the in-situ measured data.

3.1 Lake Biwa Satellite-Derived Chlorophyll-a

When compared with Landsat-8, the Lake Biwa chlorophyll-a derived from the Sentinel-2A gives a good fit ($R^2 = 0.76$) with the in-situ measured data. The downloaded images of the two satellite sensors has the difference of one day, therefore the estimated chlorophyll-a values were not significantly different in the two sensor images. The chlorophyll-a map generated using the Landsat-8 and Sentinel-2A is shown in Figure 2. In both sensors, the estimated chlorophyll-a in the south basin of Lake Biwa was relatively higher than the north basin.

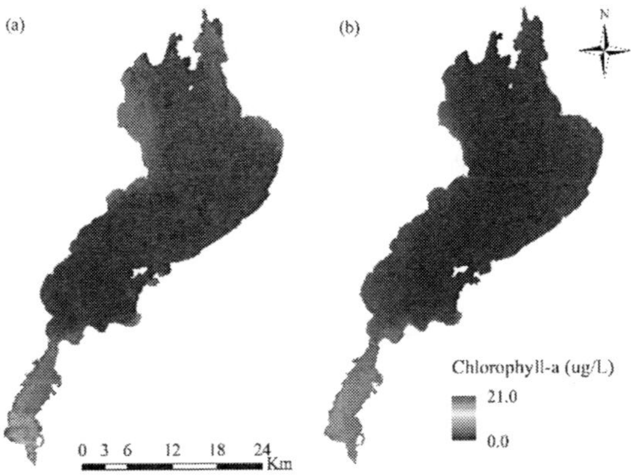

Figure 2. Estimated chlorophyll-a for Lake Biwa using two different satellite sensors: (a) Landsat-8 image and (b) Sentinel-2A image for August 2018. (Note: ug/L represents microgram per liter).

Of 18 sampling locations in Lake Biwa, eight locations were used to validate the Landsat-8 and Sentinel-2A derived chlorophyll-a results. Sentinel-2A appears to give a good fit with the in-situ measured data as compared to Landsat-8 (R^2 = 0.65). The observed chlorophyll-a and chlorophyll-a derived from Landsat-8 and Sentinel-2A for Lake Biwa was evaluated as shown in Figure 3 and Figure 4, respectively.

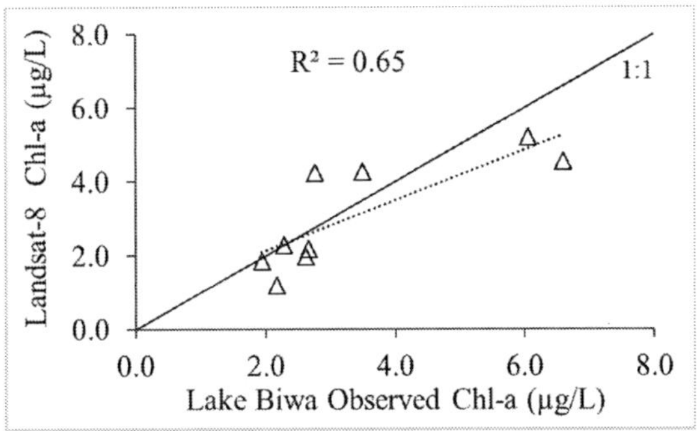

Figure 3. Observed and Landsat-8 derived chlorophyll-a for Lake Biwa (3 August 2018).

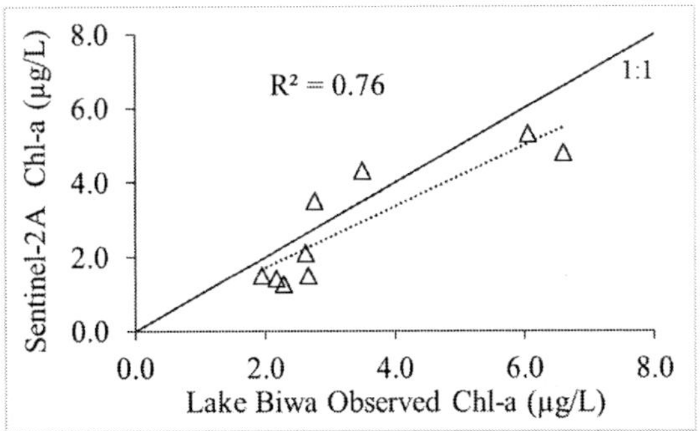

Figure 4. Observed and Sentinel-2A derived chlorophyll-a for Lake Biwa (4 August 2018).

With the increase in chlorophyll-a concentration both the sensors show slight underestimation. The maximum chlorophyll-a was noted near the shoreline of the south basin which is 21 μg/L by Landsat-8 and 19 μg/L by Sentinel-2A. Based on the observed data, the root mean square error (RMSE) was estimated for both the sensors. The RMSE for the Sentinel-2A was 0.97 μg/L and Landsat-8 was 1.02 μg/L, as given in the Table1.

Chlorophyll-a (μg/L)	Observed	S-2A	L-8	N
Min	1.9	1.3	1.2	
Max	6.6	5.3	5.2	8
Average	3.4	2.9	3.1	
RMSE		0.97	1.02	

Table 1. Accuracy assessment of satellite-derived chlorophyll-a in μg/L for the Lake Biwa using in-situ data. (S-2A = Sentinel-2A; L-8 = Landsat-8, N= number of validation points).

3.2 Wakasa Bay Satellite-Derived Chlorophyll-a

In Wakasa Bay, which receives the inflow from the Yura River, the maximum chlorophyll-a concentration estimated from the two sensors was 8.5 μg/L, near the river mouth. The satellite-derived chlorophyll-a using Landsat-8 and Sentinel-2A is shown in Figure 5.

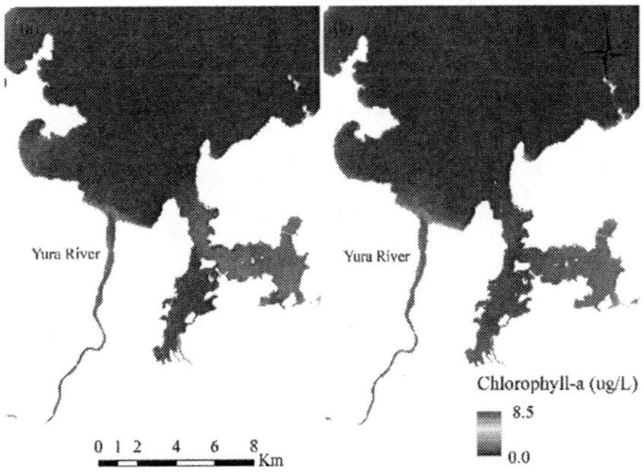

Figure 5. Estimated chlorophyll-a for Wakasa Bay coastal zone (Yura Estuary) using two different satellite sensors: (a) Landsat-8 image and (b) Sentinel-2A image for August 2018. (Note: ug/L represents microgram per liter).

No significant visible difference was observed between the satellite-derived chlorophyll-a of the two sensors. Of 21 sampling locations, 12 locations were used for the validation of the results from two satellite-sensors. Both sensors, appears to give the best result with R^2 = 0.81 (Landsat-8) and R^2 = 0.83 (Sentinel-2A). The results of the chlorophyll-a derived from Landsat-8 and Sentinel-2A for Wakasa Bay is shown in Figure 6 and Figure 7, respectively.

https://doi.org/10.5194/isprs-archives-XLII-3-W7-77-2019

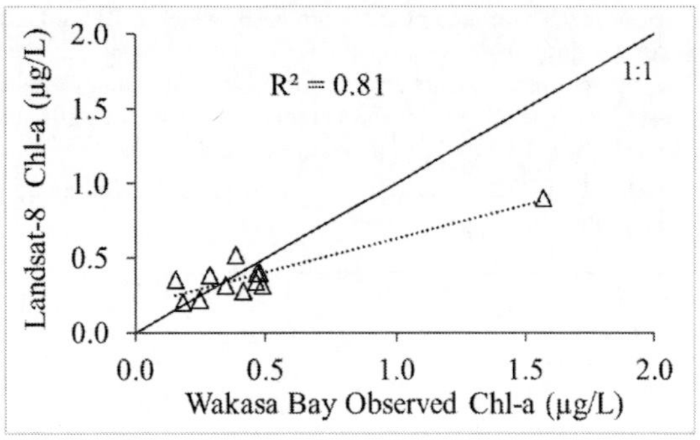

Figure 6. Observed and Landsat-8 derived chlorophyll-a for Wakasa Bay (3 August 2018).

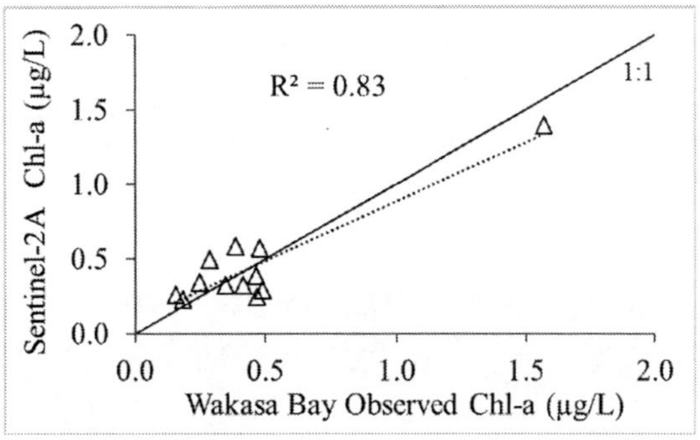

Figure 7. Observed and Sentinel-2A derived chlorophyll-a for Wakasa Bay (4 August 2018).

Although the results showed a good fit in Wakasa Bay, the underestimation of chlorophyll-a was observed particularly in the Landsat-8 results with the increase in the chlorophyll-a concentration. The underestimation in the Landsat-8 could be due to the differences in the spatial resolution. For instance, Landsat-8 has 30 m resolution thus each pixel represents 900 m^2 of the area however the estimated results were compared with the in-situ measured point data of chlorophyll-a. Consequently, a slight difference in the results was obtained. Furthermore, Yura river is a major river inflow into the Wakasa Bay (Antonio et al., 2010). Thus, a possible difference in chlorophyll-a concentration due to the river flow was also expected in both the sensors. Obtained results for the Bay area indicates slightly better applicability of the Sentinel-2A for chlorophyll-a estimation. The RMSE calculated for the satellite-derived chlorophyll-a of Wakasa Bay is given in Table 2.

Chlorophyll-a (µg/L)	Observed	S-2A	L8	N
Min	0.16	0.22	0.20	12
Max	1.57	1.39	0.90	
Average	0.46	0.45	0.39	
RMSE		0.14	0.22	

Table 2. Accuracy assessment of satellite-derived chlorophyll-a for the Wakasa Bay using in-situ data. (S-2A = Sentinel-2A; L-8 = Landsat-8, N= number of validation points).

4. CONCLUSION

In this study, efforts being made to test the suitability of the two different sensors Landsat-8 and Sentinel-2A for chlorophyll-a estimation in freshwater (Lake Biwa) and coastal water (Wakasa Bay - coast of Yura River). Moreover, the applicability of the Spectral Decomposition Algorithm to evaluate the chlorophyll concentration was also examined for coastal water along with the freshwater body. The algorithm was successfully applied to both the optically complex water bodies giving the root mean square error of < 2 µg/L. Relatively higher chlorophyll-a concentration obtained through the satellite sensors for both the water bodies particularly near the shorelines and the river mouth (when compared with the inshore areas) also indicates the significant nutrient load through the river runoff and from the surrounding land use activities. A lower root mean square error < 1 µg/L suggests a better performance of Sentinel-2A for both the coastal and freshwater body than the Landsat-8. The medium resolution (i.e., 30 m) of Landsat-8 and discharge from the large river (in the coastal area) were suspected to be the cause of underestimation of the chlorophyll-a. Information of variation in climatic factors (i.e., rainfall) and river runoff are also essential as it could affect the water quality of a given water body.

ACKNOWLEDGEMENTS

This study is the part of the CoHHo project (Connectivity of Hills, Humans, and Oceans) supported by the Nippon Foundation. We are thankful for the help provided by Lake Biwa Environmental Research Institute (LBERI, Otsu) for data collection and laboratory support.

REFERENCES

Antonio, E.S., Kasai, A., Ueno, M., Won, N. il, Ishihi, Y., Yokoyama, H., Yamashita, Y., 2010. Spatial variation in organic matter utilization by benthic communities from Yura River-Estuary to offshore of Tango Sea, Japan. Estuar. Coast. Shelf Sci. 86, 107–117. https://doi.org/10.1016/j.ecss.2009.10.020.

Bhagowati, B., Ahamad, K.U., 2018. A review on lake eutrophication dynamics and recent developments in lake modeling. Ecohydrol. Hydrobiol. https://doi.org/10.1016/j.ecohyd.2018.03.002.

Blondeau-Patissier, D., Gower, J.F.R., Dekker, A.G., Phinn, S.R., Brando, V.E., 2014. A review of ocean color remote sensing methods and statistical techniques for the detection, mapping and analysis of phytoplankton blooms in coastal and open oceans. Prog. Oceanogr. 123, 23–144. https://doi.org/10.1016/j.pocean.2013.12.008.

Carpenter, S.R., Caraco, N.F., Correll, D.L., Howarth, R.W., Sharpley, A.N., Smith, V.H., 1998. Vibrational spectra of crystalline n-paraffins. II. Intermolecular effects. Ecol. Appl. 8, 559–568. https://doi.org/10.1016/0022-2852(61)90347-2.

Cheng, C., Wei, Y., Sun, X., Zhou, Y., 2013. Estimation of chlorophyll-a concentration in Turbid Lake using spectral smoothing and derivative analysis. Int. J. Environ. Res. Public Health 10, 2979–2994. https://doi.org/10.3390/ijerph10072979.

Cloern, J.E., Jassby, A.D., 2010. Patterns and scales of phytoplankton variability in estuarine-coastal ecosystems.

Estuaries and Coasts 33, 230–241. https://doi.org/10.1007/s12237-009-9195-3.

Cullen, J.J., 1982. The Deep Chlorophyll Maximum: Comparing Vertical Profiles of Chlorophyll a. Can. J. Fish. Aquat. Sci. 39, 791–803. https://doi.org/https://doi.org/10.1139/f82-108.

Dekker, A.G., Peters, S.W.M., 1993. The use of the Thematic Mapper for the analysis of eutrophic lakes: a case study in the Netherlands. Int. J. Remote Sens. 14, 799–821.

Diaz, R.J., 2009. Spreading Dead Zones and Consequences for Marine Ecosystems. Science (80) 926, 926–930. https://doi.org/10.1126/science.1156401.

Fredston-Hermann, A., Brown, C.J., Albert, S., Klein, C.J., Mangubhai, S., Nelson, J.L., Teneva, L., Wenger, A., Gaines, S.D., Halpern, B.S., 2016. Where Does River Runoff Matter for Coastal Marine Conservation? Front. Mar. Sci. 3, 1–10. https://doi.org/https://doi.org/10.1890/10510761(1998)008[0559:NPOSWW]2.0.CO;2.

Ha, N.T.T., Thao, N.T.P., Koike, K., Nhuan, M.T., 2017. Selecting the Best Band Ratio to Estimate Chlorophyll-a Concentration in a Tropical Freshwater Lake Using Sentinel 2A Images from a Case Study of Lake Ba Be (Northern Vietnam). ISPRS Int. J. Geo-Information 6, 290. https://doi.org/10.3390/ijgi6090290.

Ishikawa, K., Kumagai, M., Vincent, W.F., Tsujimura, S., Nakahara, H., 2002. Transport and accumulation of bloom-forming cyanobacteria in a large, mid-latitude lake: The gyre-Microcystis hypothesis. Limnology 3, 87–96. https://doi.org/10.1007/s102010200010.

Kändler, M., Blechinger, K., Seidler, C., Pavlů, V., Šanda, M., Dostál, T., Krása, J., Vitvar, T., Štich, M., 2017. Impact of land use on water quality in the upper Nisa catchment in the Czech Republic and in Germany. Sci. Total Environ. 586, 1316–1325. https://doi.org/10.1016/j.scitotenv.2016.10.221.

Kasai, A., Kurikawa, Y., Ueno, M., Robert, D., Yamashita, Y., 2010. Salt-wedge intrusion of seawater and its implication for phytoplankton dynamics in the Yura Estuary, Japan. Estuar. Coast. Shelf Sci. 86, 408–414. https://doi.org/10.1016/j.ecss.2009.06.001.

Le, C., Hu, C., Cannizzaro, J., English, D., Muller-Karger, F., Lee, Z., 2013. Evaluation of chlorophyll-a remote sensing algorithms for an optically complex estuary. Remote Sens. Environ. 129, 75–89. https://doi.org/10.1016/j.rse.2012.11.001.

Le, C., Li, Y., Zha, Y., Sun, D., Huang, C., Lu, H., 2009. A four-band semi-analytical model for estimating chlorophyll a in highly turbid lakes: The case of Taihu Lake, China. Remote Sens. Environ. 113, 1175–1182. https://doi.org/10.1016/j.rse.2009.02.005.

Lee, Z., Carder, K.L., Mobley, C.D., Steward, R.G., Patch, J.S., 1998. Hyperspectral remote sensing for shallow waters I A semianalytical model. Appl. Opt. 37, 6329. https://doi.org/10.1364/AO.37.006329.

Lintern, A., Webb, J.A., Ryu, D., Liu, S., Bende-Michl, U., Waters, D., Leahy, P., Wilson, P., Western, A.W., 2017. Key factors influencing differences in stream water quality across space. Wiley Interdiscip. Rev. Water 5, e1260. https://doi.org/10.1002/wat2.1260.

Matsushita, B., Yang, W., Chang, P., Yang, F., Fukushima, T., 2012. A simple method for distinguishing global Case-1 and Case-2 waters using SeaWiFS measurements. ISPRS J. Photogramm. Remote Sens. 69, 74–87. https://doi.org/10.1016/j.isprsjprs.2012.02.008.

Mobley, C., Stramski, D., Bissett, P., Boss, E., 2004. Optical Modeling of Ocean Waters: Is the Case 1 - Case 2 Classification Still Useful? Oceanography 17, 60–67. https://doi.org/10.5670/oceanog.2004.48.

Morel, A., 1988. Optical modeling of the upper ocean in relation to its biogenous matter content (case I waters). J. Geophys. Res. 93, 10749. https://doi.org/10.1029/JC093iC09p10749.

Oyama, Y., Matsushita, B., Fukushima, T., Matsushige, K., Imai, A., 2009. Application of spectral decomposition algorithm for mapping water quality in a turbid lake (Lake Kasumigaura, Japan) from Landsat TM data. ISPRS J. Photogramm. Remote Sens. 64, 73–85. https://doi.org/10.1016/j.isprsjprs.2008.04.005.

Reinart, A., Kutser, T., 2006. Comparison of different satellite sensors in detecting cyanobacterial bloom events in the Baltic Sea. Remote Sens. Environ. 102, 74–85. https://doi.org/10.1016/j.rse.2006.02.013.

Shi, K., Li, Y., Li, L., Lu, H., Song, K., Liu, Z., Xu, Y., Li, Z., 2013. Remote chlorophyll-a estimates for inland waters based on a cluster-based classification. Sci. Total Environ. 444, 1–15. https://doi.org/10.1016/j.scitotenv.2012.11.058.

Silva, S.V.S., Dias, A.H.C., Dutra, E.S., Pavanin, A.L., Morelli, S., Pereira, B.B., 2016. The impact of water pollution on fish species in southeast region of Goiás, Brazil. J. Toxicol. Environ. Health 79, 8–16. https://doi.org/https://doi.org/10.1080/15287394.2015.1099484.

Toming, K., Kutser, T., Laas, A., Sepp, M., Paavel, B., Nõges, T., 2016. First experiences in mapping lakewater quality parameters with sentinel-2 MSI imagery. Remote Sens. 8, 1–14. https://doi.org/10.3390/rs8080640.

USGS, n.d. United States Geological Survey (USGS) EarthExplorer [WWW Document]. URL https://earthexplorer.usgs.gov/.

Wozniak, M., Bradtke, K.M., Krezel, A., 2014. Comparison of satellite chlorophyll a algorithms for the Baltic Sea. J. Appl. Remote Sens. 8, 083605. https://doi.org/10.1117/1.JRS.8.083605.

Yadav, S., Yoneda, M., Tamura, M., Susaki, J., Ishikawa, K., Yamashiki, Y., 2017. A satellite-based assessment of the distribution and biomass of submerged aquatic vegetation in the optically shallow basin of Lake Biwa. Remote Sens. 9, 966. https://doi.org/10.3390/rs9090966.

Zhang, L., Xu, E.G., Li, Y., Liu, H., Vidal-Dorsch, D.E., Giesy, J.P., 2018. Ecological risks posed by ammonia nitrogen (AN) and un-ionized ammonia (NH<inf>3</inf>) in seven major river systems of China. Chemosphere 202. https://doi.org/10.1016/j.chemosphere.2018.03.098

ASSESSMENT OF CLOUD COVER CHARACTERISTICS OVER CALIBRATION TEST SITES USING MODIS CLOUD MASK PRODUCTS

H. Yamamoto[1], S. Tsuchida[1]

[1] Geological Survey of Japan, AIST - (hirokazu.yamamoto, s.tsuchida)@aist.go.jp

Commission III, WG III/10

KEY WORDS: Vicarious calibration, Calibration sites, Railroad valley playa, Clear sky ratio, MODIS cloud mask product, MOD35_L2, MYD35_L2

ABSTRACT:

Spaceborne remotely sensed data can provide the spatial, spectral, and temporal coverage for earth monitoring. Radiometric calibration is indispensable for the accurate higher level products derived from satellite optical sensors, and the long-term radiometric calibration has a crucial role. Especially, vicarious calibration is currently the most accurate method to conduct the radiometric calibration of satellite optical sensors. Various organization and projects have been conducting the vicarious calibration experiments for satellite optical sensors. Railroad valley playa is the most popular calibration site, and we have been also acquiring the vicarious calibration data for Terra ASTER there from the past. Recently, RadCalNet (Radiometric Calibration Network) is open to the public, which is an initiative of the CEOS WGCV. Railroad valley is also one of the RadCalNet sites, and our calibration site is located at the adjacent area. Understanding the cloud cover characteristics over calibration test sites is very important for vicarious calibration and automated facilities measurement. This research evaluates the clear sky ratio for Railroad valley playa using Terra an Aqua MODIS cloud mask product (MOD35_L2 and MYD35_L2) since 2000.

1. INTRODUCTION

Many researches related with environmental monitoring use data obtained by satellite optical sensors, and radiometeric accuracy is very important for deriving the biophysical parameters. Therefore, radiometric calibration has a key role in the accuracy of higher level products obtained by satellite optical sensors. Vicarious calibration is one of techniques of radiomteric calibration for satellite sensors(Slater et al., 1996).

The reflectance factor on the ground, temperature, atmospheric parameters, and so on are obtained for estimeting the top of atmosphere (TOA) reflectances when target satellite sensor flies over the specific site (Desert, Salt Lake, and so on). It is desirable to have much data obtained over the calibration sites as possible due to the repeat cycle of the satellite. Hence, recently, the Infrared and Visible Optical Sensors subgroup of the Committee on Earth Observation Satellite (CEOS) Working Group on Calibration and Validation (WGCV) set up the radiometric test sites network, the Radiometric Calibration Network (RadCalNet, https://www.radcalnet.org), which provides the TOA reflectances from 380nm to 2500nm at a 10 nm spectral sampling interval(Czapla-Myers et al., 2016).

We have conducted vicarious calibration experiments for Terra ASTER sensor(Yamaguchi et al., 1998) in the dry lake beds and salt lake every year since 1995 (Figure 2). Railroad valley playa is one of our main calibration sites, which is located at 38.505°N, 115.690°W, 1436 [m] altitude in east-central Nevada, US (Figure 1). The Radiometric Calibration Test Site (RadCaTS) at Railroad Valley, Nevada was developed by the University of Arizona, which is one of RadCalNet sites. RadCaTS is located at the adjacent area of our calibration site. Many calibration groups in various satellite mission have used Railroad valley playa as calibration site (MODIS, ASTER, GOSAT, MISR, OCO, Landsat, Setinel, and so on)(Thome et al., 2003; Kuze et al., 2011; Bruegge et al., 2014; Czapla-Myers et al., 2017).

Calibration sites need to be homogeneous and flat, and it is preferable that the characteristic of surface reflectance from visible to short-wavelength infrared region is stable. In addition to the logistics and accessibility, the weather and atmospheric conditions of the sites are also important. The objective of this research is to understand the cloud cover trend over the calibration site, Railroad valley playa, by using MODIS cloud mask (MOD35_L2 and MYD35_L2) products.

Figure 1. Location of Railroad Valley Playa

2. METHODOLOGY

We use the Terra MODIS Cloud Mask product (MOD35_L2) from 2000 to 2018, and the Aqua MODIS product (MYD35_L2) from 2002 to 2018 in this research. MOD35_L2 and MYD35_L2

Figure 2. Vicarious calibration experiment in Railroad valley playa (Jun, 2012)

are daily global Level 2 products, and provide 48-bit (6 byte) field in each 1-km pixel(Strabala, 2005, Ackerman, 2010). The version of all products we used is collection 6.1, which is the latest MODIS cloud mask product. There are four types in "Unobstructed FOV Confidence Flag" of these products, which are Cloudy, Uncertain, Probably Clear, and Confident Clear (Table 1). This research define "Confident Clear" and "Probably Clear" as clear-sky, and we simply calculatete the clear sky ratio $r(t)$ as below, We extracted the cloud flag values over our measurement sites in Railroad valley playa.

Table 1. Bit fields within first bite of the 48-bit MODIS cloud mask (MOD35_L2, MYD35_L2)

Bit field	Description	Key
0	Cloud Mask Flag	0 = Not determined 1 = Determined
2, 1	Unobstructed FOV Quality Flag	00 = Cloudy 01 = Uncertain 10 = Probably Clear 11 = Confident Clear
3	Day or Night Path	0 = Night 1 = Day
4	Sunglint Path	0 = Yes 1 = No
5	Snow/Ice Background Path	0 = Yes 1 = No
7, 6	Land or Water Path	00 = Water 01 = Coastal 10 = Desert 11 = Land

$$r(t) = \frac{N(t)}{5}. \tag{1}$$

t : Day Of Year (DOY, t=3∼ 363 or 364)
$r(t)$: Clear sky ratio at t
$N(t)$: The number of clear sky days ("Probably Clear" or "Confident Clear" in MOD35_L2 or MYD35_L2 flag) in five days($t - 2 \sim t + 2$)

All products are downloaded from the Level-1 and Atmosphere Archive & Distribution System (LAADS) Distributed Active Archive Center (DAAC), NASA, and 10,838 MOD35_L2 and 9,721 MYD35_L2 products were processed.

3. RESULTS

Figure 3 shows mean values of clear sky ratio $r(t)$ using MOD35_L2 in 2000-2018 and MYD35_L2 in 2002-2018 for each day. Figure 4 and Figure 5 depict every 5 years mean values of clear sky ratio $r(t)$ from 2004 to 2018 in the morning and afternoon, respectively. Table 2 shows the summary of clear sky ratio $r(t)$ statistics in four seasons.

Highest clear sky ratio, $r(t)$ (> 0.900) in the morning is $t = 176 \sim 261$, which means that date on which we conduct the vicarious calibration experiment over Railroad valley playa should be set to arround Jun 25∼Sep 18. On the other hand, there are no highest $r(t)$ in the afternoon, and higher $r(t)$ (> 0.850) in the afternoon is $t = 170$(Jun 19) $\sim$ 268(Sep 25). $r(t)$ of MOD35_L2 is slightly higher than MYD35_L2, that difference is approximately 6.5%. Figure 4 and Figure 5 indicate that past trend of clear sky ratio is almost same as the current trend.

Table 2. Clear sky ratio derived from MOD35_L2 (2000-2018) and MYD35_L2(2002-2018) in four seasons

Season	Product	Mean	Standard Deviation
Spring (Mar,Apr,May)	MOD35_L2	0.641	0.106
	MYD35_L2	0.584	0.090
Summer (Jun,Jul,Aug)	MOD35_L2	0.866	0.045
	MYD35_L2	0.801	0.061
Autumn (Sep,Oct,Nov)	MOD35_L2	0.670	0.192
	MYD35_L2	0.669	0.153
Spring (Dec,Jan,Feb)	MOD35_L2	0.343	0.048
	MYD35_L2	0.387	0.077

4. CONCLUSIONS

We invedtigated the clear sky ratio over Railroad valley playa, which is the vicarious calibration site for various satelite missions.

The preferable date on which conduct the vicarious calibration experiments over Railroad valley playa is from late June to late September. Experiments in the morning could have slightly advantages to afternoon experiments.

Although vicarious calibration is very accurate calibration techniques for satellite optical sensors, this needs various resources (human, time, cost, and so on). The results in this research is very useful information for vicarious calibration plan. And, the methodology of clear sky ratio using MODIS cloud mask products in this study could be also applicable to assessnebt of validation sites.

This contribution has been peer-reviewed.
https://doi.org/10.5194/isprs-archives-XLII-3-W7-83-2019 | © Authors 2019. CC BY 4.0 License.

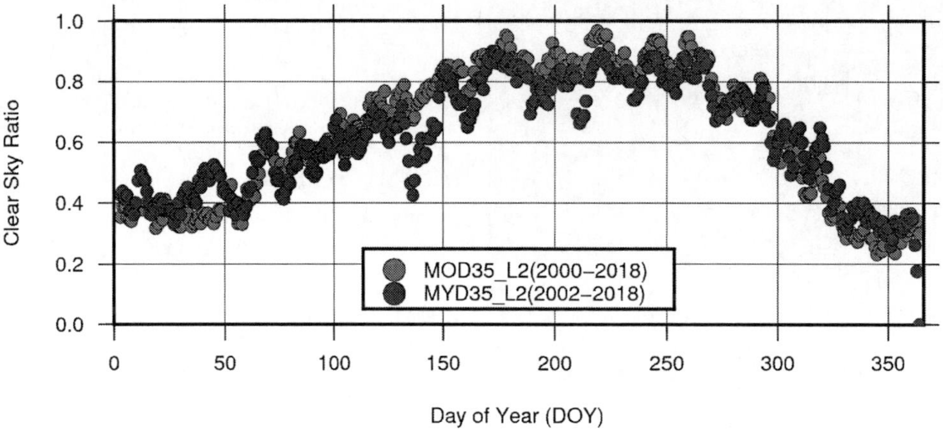

Figure 3. Mean value of clear sky ratio $r(t)$ over Railroad Valley

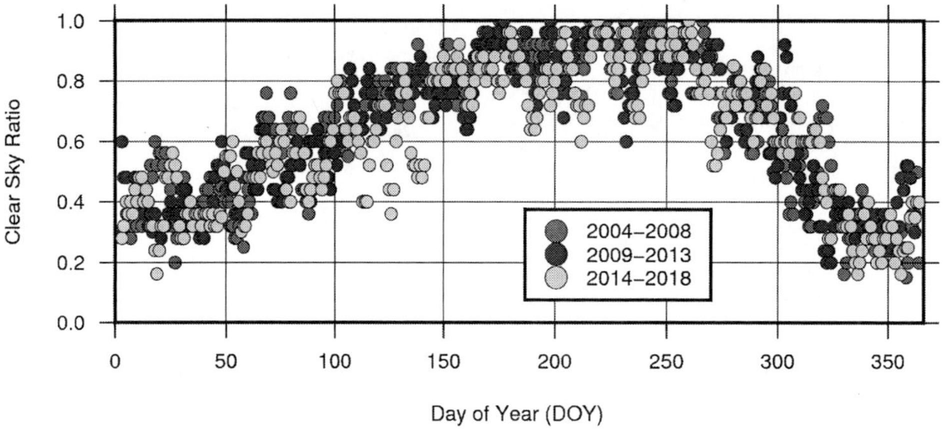

Figure 4. Mean value for a five year period of clear sky ratio $r(t)$ over Railroad Valley (MOD35_L2)

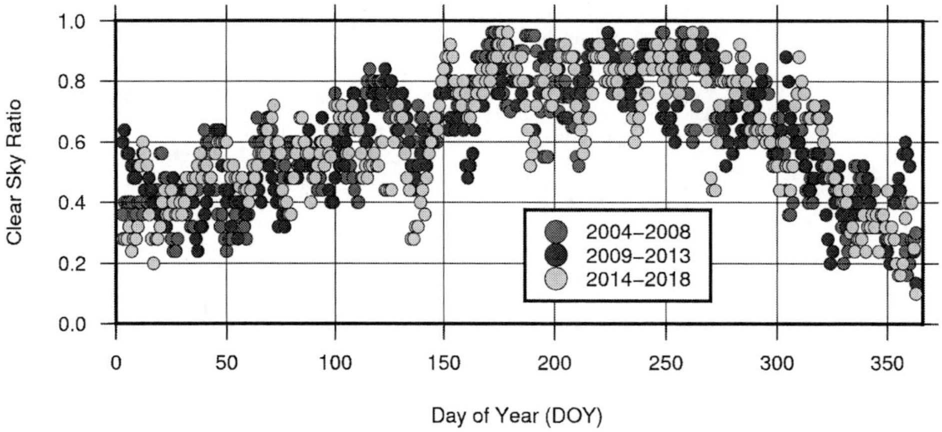

Figure 5. Mean value for a five year period of clear sky ratio $r(t)$ over Railroad Valley (MYD35_L2)

ACKNOWLEDGEMENTS

The Terra and Aqua MODIS Atmosphere L2 Cloud Mask Products were acquired from the Level-1 and Atmosphere Archive & Distribution System (LAADS) Distributed Active Archive Center (DAAC), located in the Goddard Space Flight Center in Greenbelt, Maryland (https://ladsweb.nascom.nasa.gov/).

REFERENCES

Ackerman, S., 2010. Discriminating clear-sky from cloud with modis. *Algorithm Theoretical Basis Document MOD35 Version 6.1*.

Bruegge, C. J., Val, S., Diner, D. J., Jovanovic, V., Gray, E., Di Girolamo, L. and Zhao, G., 2014. Radiometric stability of the multi-angle imaging spectroradiometer (misr) following 15 years on-orbit. In: *Earth Observing Systems XIX*, Vol. 9218, International Society for Optics and Photonics, p. 92180N.

Czapla-Myers, J., Bouvet, M. and Wenny, B., 2016. The radiometric calibration network (radcalnet): a global calibration and validation test site network. In: *AGU Fall Meeting Abstracts*.

Czapla-Myers, J., McCorkel, J., Anderson, N. and Biggar, S., 2017. Earth-observing satellite intercomparison using the radiometric calibration test site at railroad valley. *Journal of Applied Remote Sensing* 12(1), pp. 012004.

Kuze, A., O'Brien, D. M., Taylor, T. E., Day, J. O., O'Dell, C. W., Kataoka, F., Yoshida, M., Mitomi, Y., Bruegge, C. J., Pollock, H. et al., 2011. Vicarious calibration of the gosat sensors using the railroad valley desert playa. *IEEE Transactions on Geoscience and Remote Sensing* 49(5), pp. 1781–1795.

Slater, P. N., Biggar, S. F., Thome, K. J., Gellman, D. I. and Spyak, P. R., 1996. Vicarious radiometric calibrations of eos sensors. *Journal of Atmospheric and Oceanic Technology* 13(2), pp. 349–359.

Strabala, K. I., 2005. *MODIS cloud mask user's guide*. University of Wisconsin–Madison.

Thome, K. J., Czapla-Myers, J. S. and Biggar, S. F., 2003. Vicarious calibration of aqua and terra modis. In: *Earth Observing Systems VIII*, Vol. 5151, International Society for Optics and Photonics, pp. 395–406.

Yamaguchi, Y., Kahle, A., Tsu, H., Kawakami, T. and Pniel, M., 1998. Overview of Advanced Spaceborne Thermal Emission and ReflectionRadiometer (ASTER). *IEEE Transactions on Geoscience and Remote Sensing* 36(4), pp. 1062–1071.

Revised Jan 2019

https://doi.org/10.5194/isprs-archives-XLII-3-W7-83-2019

AUTHOR INDEX

Abe, T. 64
Balzter, H. 53
Barbosa-Herrera, A. P. 53
Barreto-Silva, J. S. 53
Bispo, P. C. 53
Cabrera, E. 53
Capachero, C. 53
Chen, X. 1
Cho, K. 5, 43
Dewayani, S. 68
Duong, N. D. 10
Galindo, G. 53
Gou, Y. 53
Hidayat, D. 68
Hongo, C. 68
Ishii, Y. 17
Ishikawa, K. 72
Iwao, K. 17
Jamali, A. 23
Khairiah, R. N. 31
Kinoshita, T. 17
Kitagami, T. 68
Kumagai, K. 38
Lee, H. 1
Lee, M. 1
Louis, V. 53
Lozano, P. 53
Matsumoto, M. 43
Miyao, K. 5
Moreno, L. M. 53
Motohashi, K. 68
Naoki, K. 5, 43
Ohki, M. 64
Pacheco-Pascagaza, A. M. 53
Pachon-Cendales, I. P. 53
Pantanahiran, W. 47
Phillips-Bernal, J. F. 53
Phuong, N. M. 10
Prasetyo, L. B. 31
Rikimaru, A. 57
Roberts, J. 53
Rodríguez-Veiga, P. 53
Sakata, K. 57
Salinas, N. R. 53
Setiawan, Y. 31
Susaki, J. 72
Tadono, T. 64
Thi, N. B. 10
Tjahjono, B. 68
Tsuchida, S. 78
Uematsu, H. 38
Vergara, L. 53
Wakabayashi, H. 43, 68
Yadav, S. 72
Yamamoto, H. 78
Yamashiki, Y. 72
Yamashita, Y. 72
Yasunaga, Y. 57
Yoshimura, M. 43
Zuluaga, A. C. 53